Crunch Time

Crunch Time

How Everyday Life is Killing the Future

Adrian Monck and Mike Hanley

www.crunchtime.info

ICON BOOKS

This edition published in 2007 by
Icon Books Ltd, The Old Dairy, Brook Road,
Thriplow, Cambridge SG8 7RG
email: info@iconbooks.co.uk
www.iconbooks.co.uk

Originally published in Sydney in 2004 by Allen & Unwin

This edition is not to be published by Allen & Unwin in Australia

Sold in the UK, Europe, South Africa and Asia
by Faber & Faber Ltd, 3 Queen Square,
London WC1N 3AU
or their agents

Distributed in the UK, Europe, South Africa and Asia
by TBS Ltd, TBS Distribution Centre, Colchester Road,
Frating Green, Colchester CO7 7DW

Distributed in Canada by
Penguin Books Canada,
90 Eglinton Avenue East, Suite 700,
Toronto, Ontario M4P 2YE

ISBN 10: 1-84046-801-7
ISBN 13: 978-1840468-01-4

Typesetting by Wayzgoose

Printed and bound in the UK by
Clays of Bungay

Contents

Adrian says: 'This is for Linda,
Ella and Ethan'

Mike says: 'This is for Claire, Max,
Joel and now Sally'

Foreword

Some days it seems that before we've even finished eating breakfast we've committed crimes against the environment, trampled over social justice or ignored some looming catastrophe.

Are the things that we do routinely – commuting to work, buying toys for our kids, preparing meals, having a cup of coffee – conspiring to destroy our collective future?

Our time – Crunch Time – is saturated with distractions. Wherever we turn there are marvellous spectacles, intriguing technological magic, transforming art and transfixing low culture. But you don't have to travel far to confront the ugly flip side of our modern, globalised world.

Click on a website or turn on the TV and out leap stories that define this extraordinary epoch – climate change, the rise of global terrorism, energy security, global flu threats, political conflagration in the Middle East.

It's easy to be overwhelmed by the sheer scale of the issues that confront us, and daunted by the way they seem to grip every detail of daily living – our choice of lunch as much as our choice of career.

To begin to understand them you don't need a guidebook, you need a starting point – and that's where Adrian and Mike come in. That start doesn't come with glib solutions. Adrian and Mike don't always see eye to eye and, like me, you'll probably find yourself joining in the argument as often as agreeing.

But that's the start. Stepping into the big debates with

the same confidence as political, military and business leaders is what it's all about. That's where this book comes in – you might just find your way into the arena here.

Crunch Time is our time. It's where we try to figure out how to bring up children in a world we don't fully understand, how to make sense of our working lives, and how to get along with the other people and cultures with whom we share an ever-shrinking planet.

And we might be able to defend the future from the people who most threaten it – ourselves.

Kirsty Young

INTRODUCTION

The Sky is Falling

The world is in haste, and nears its end

Wulfstan II, Archbishop of York 1014

This is not a book about the world coming to an end. But it is a book about a world that is coming to an end. Our world. It's Crunch Time for the world as we know it, but we don't feel fine. We think that either the human race will get it right over the next 50 years, or it will be game over. It's up to us to determine which it will be.

The collective decisions we make now will directly influence the state of the world in, say, 50 years' time. For simplicity's sake, assume that there are two possible poles of reality for 1 January 2050. In the first scenario, imagine we get it right. In this scenario, the sun will rise on a world where many billions of people enjoy lives of unfettered economic, social and political liberty. They pursue their dreams aided by whiz-bang technology, the likes of which we have yet to even imagine, while the environment is safeguarded by sustainable technology against the destruction the human race would otherwise wreak upon it, and its powers of regeneration and evolution are undimmed. The average human life is relatively long and fulfilled.

On the other hand, imagine we get it wrong. In many ways, this is the easier picture to draw, assisted as we are every day by television and the newspapers. At dawn on

1 January 2050, the sun will rise on a world that contains no people at all – the human race having extinguished itself or fled from the planet some years before. All that's left is a damaged web of life struggling to overcome the tears in its fabric, which were both caused by, and the cause of, the extinction of its most destructive species ever.

The 20th century and all its attendant horrors raised the stakes for the future way beyond anything that had come before. We believe that within our lifetimes it will be determined whether the human race can secure a permanent place at the universe's table or whether our evolutionary chips are to be cashed in. It's now double or nothing.

It's by no means a no-brainer, though, and things are not all grim. If you were to bet on humankind seeing out the next 50 years, you might point to the many positive developments in human endeavour the last century saw, including the unprecedented increase in health, longevity and prosperity across much of the world, from America and the rest of the rich world to Asia and beyond – with strong indications of more to come; the fall of totalitarian regimes across the globe, from the fascist military dictators who ruled much of Latin America through to the communist regimes of the Eastern bloc; fast, cheap international travel; the plummeting cost of communications, telephone, internet, television and the rest; rising education and literacy standards; great leaps of scientific understanding, and a corresponding ability to manipulate the world to our own ends; the promise of discoveries in medicine and genetics; and the general resilience demonstrated by the human species when struck by both natural and self-created disasters. There are many reasons to think that the world in 2050 will, in almost every way, be a better place to be born into.

On the other hand, there are many reasons to believe that we won't be around to collect on the bet. Nuclear weapons may have grown old but their destructive power has not decreased, while their number and ownership have multiplied. Meanwhile, human numbers are set to rise by 50 per cent over the next half-century. By 2050, some 9 billion souls will press in on a world of scarcity. The scientific and technological progress that we point to with such awe has a terrifying flip side: look at the ructions caused by genetically modified foods, cloning and the human genome project to name just three. And the environment is a rich source of apocalyptic gloom, from global warming, threats to water supplies, increasing incidences of natural catastrophes, disappearing species and deforestation to local pollution and environmental poisoning. Resources we rely on, oil for instance, are running down. AIDS, malaria, tuberculosis and a host of other bacteria and viruses show at least as great a talent for survival as we do. New diseases like SARS and H5N1 (bird flu) and unknown mutations of these threaten to create plagues that might hearken back to the Black Death. Wars big and small proliferate. Meanwhile, the poor remain poor, the hungry remain malnourished and don't even ask about the meek.

If you were of a pessimistic persuasion, you could argue strongly that we're our own worst enemy and our race is all but run.

Is there a feeling of déjà vu here? Surely, the same poles of possibility – one optimistic and the other pessimistic – existed for anyone looking forward to the millennium in the middle of last century? Then, the gloomy prognosis was of a Cold War threatening nuclear apocalypse, or else a world sunk in a long night of totalitarianism; whilst the cheerier outlook was to view the United States' post-war consumer boom as final delivery of the founding fathers' promise of 'the pursuit of happiness'. Surely the same two poles were visible to Winston Churchill when he gazed out upon the future? Or, for that matter, to anyone gazing into the future, from Jesus of Nazareth through to Stephen Hawking.

Perhaps people always think that their times are more critical than others and it's a basic human trait to believe that our own part in historical development is more important than those who have come before. In the end, the world in 2000 was a mixture of both ends of the spectrum and everything in between. This may be what's in store for us in 2050. We might hope so, but that doesn't absolve us of the responsibility these times have placed on our shoulders.

Doomsaying is hardly a new trend. In what we call the Dark Ages, the then Archbishop of York, Wulfstan II, surveyed an England ravaged by Viking invasions, beset by disorder, famine and plague, and concluded that, yes, the world was swiftly drawing to a close. Wulfstan was not a gloomy hermit. He was a major figure in one of the most powerful international organisations of his time, a politician and leading government adviser, and a writer and intellectual. Wulfstan held sway over massive agricultural estates, went on diplomatic missions, and got involved in defence policy and diplomacy and the odd refugee crisis. His conviction that the world was fast coming to its end was reached from a position near the summit of global affairs.

At the time, the rich endowed monasteries as an insurance policy for their souls, and joining the Church was a way of future-proofing your fate. The churches that dot the European countryside are a reminder of just how much energy Wulfstan and his contemporaries put into allaying their own feelings of foreboding. The world didn't end but Wulfstan, and the world in which he lived, did.

Civilised societies have not stopped facing crises since. But this book's thesis is that, despite everything that has come before, we are the custodians of a world, and the way that we're living in it is threatening its destruction. And building any number of churches isn't going to help.

Saving civilisation

Are there rules and lessons we can draw on to give us an inkling of the future? Some people think so. One of them is Sid Meier. Here's his take on the building blocks of the present:

> Diplomacy, cultural and economic strength, resources, technological progress, military strength and strategy, the natural environment, all these things combine in a great maelstrom that contributes to and takes away from human development. And this happens at both the level of the individual civilization and with the progress of the human race as a whole.

Meier isn't a politician or an economist. He's a computer game designer and, in that quirky and cliquey world, something of a legend. He is one of the few men who can boost sales of a product simply by attaching his name to it, and is famous for inventing the strategy game for the computer. Meier's games are easy to spot: *Sid Meier's Pirates*, *Sid Meier's Railroads*, *Sid Meier's Alpha Centauri*. In each of these, players make decisions and deploy resources in order to pursue a wider goal – from the seemingly banal process of managing and building a golf course in *Sid Meier's SimGolf*, to winning the Battle of Gettysburg in *Sid Meier's Civil War*.

Meier's greatest invention is civilisation itself; naturally, *Sid Meier's Civilization*. The game is a sweeping, epic journey that takes players from 3,000 BC to the near future. Starting as tribal settlers, players deploy armies, build cities, cultural communities, food supplies and trade links, establish diplomatic relationships with other civilisations, discover technologies, and experience war, victory and defeat. It's pretty ambitious for a computer game, modelling, for the purposes of entertainment, the whole of human history and development. The game is now in its fourth incarnation, *Civilization IV*, and has sold in the millions since it was first launched back in 1990.

It's a popular game because, as Meier says, it 'provides a general sense of "history is cool and interesting"'. Players encounter historical landmarks and people that even school dropouts have heard of: the Great Wall, the Pyramids, Julius Caesar, Queen Elizabeth I.

You get to choose whether you want to be the Babylonians or the Zulus, the Chinese or the Scots, or whoever. But *Civilization* works at a higher level too. Within your civilisation you are all-powerful and can make all sorts of decisions spanning millennia: what transport links to create, whether to build an opera house or a grain store, whether to invade a neighbour or not. But at the same time, everything you do has consequences – building an opera house is great if your people have plenty to eat and need entertaining, but pointless if they don't – so your power is limited by the world around you. The same is true in the real world. No point to the United States invading Iraq if doing so will alienate global public opinion and destroy existing diplomatic relationships, not to mention causing massive civil unrest in that country in later rounds of the game. Perhaps that's a bad example, but you get the idea.

Nonetheless, the players of the game and the way they play make a real difference. They can choose to be peace-loving civilisations, or warlike ones, and the tone of the game is dictated by that choice. Peaceful civilisations tend to spend a lot of time building cultural icons, and that wins them kudos across the globe. Warlike civilisations spend a lot of time up to their eyeballs in blood.

In attempting to encompass the whole of human development and experience, *Civilization* is forced to take a philosophical position on trade-offs and rewards, on fair play and just deserts. Luck plays a part but in the main, battles are won and lost, technologies discovered and culture developed depending on the player's underlying strength and strategy. At the same time, both timing and the broader global scene contribute to your civilisation's future.

There are three basic ways the game can end: Wagnerian *Götterdämmerung* – conquer or be conquered; run out of time a couple of hundred years hence; or escape skywards and colonise the stars. In the tightly plotted world that Meier put together, the end goals for humanity are fairly limited, but the decisions that you make as a player, in combination with the natural laws that govern the game, determine which of these finales will come about.

Real life, as we all know too well, is neither so prescriptive, nor so easy. *Civilization* players have the advantage of rules, cheats and strategy advice. We, the players in the real human race, are neither so well briefed, nor so powerful, nor so easy to judge. But there is an increasing sense that, for modern culture and society, and for *Homo sapiens* as a species, we're facing critical decisions that could have a far greater impact than simply 'game over'.

This is the feeling that motivated us to write this book. Adrian works in London and Mike in Sydney, but despite the geographical inconvenience – an eleven-hour time difference, for starters – we have a lot in common. Mostly we share superficial things: young kids, both journalists, like a drink, etc., but we also share a conviction that the way we live is part of a process of destruction: a process beyond the unintentional consequences of capitalism, and beyond the pursuit of happiness. More than that, we are curious about how to live a half-decent life in a world that must be experienced and acted on with barely a moment for understanding it, both for our own sake and that of the children who arrived on the scene. In essence, our kids snapped us out of our youthful self-obsession, as we abruptly found – much to our chagrin – it's not all about us.

Crunch Time is the result of our investigations.

The news

Global warming, terror, poverty, crime, environmental catastrophes, war, asylum seekers, greed, fraud, bankruptcy, pandemics, polemics. Enough to put you off your cornflakes, really.

Apparently, says the news, we are the cause of the seventh (or is it the eighth?) great extinction – entire species are dying out in the thousands – dropping like flies, so to speak. Apparently, according to the RSS feeds, with genetic splicing, molecular engineering, neuroscience and stuff, we're heading into a 'post-human' future where the line between human, robot, computer and machine will blur ever fuzzier. Apparently, think the think tanks, in the decades to come there will be global water shortages, and we're going to be at each other's throats literally dying for a drink.

It goes on and on. The wider world, it seems, is becoming an ever more terrifying place.

Meanwhile, on the home front, things aren't getting any easier. Families, communities, neighbourhoods: falling apart. Divorced, single-parented, blended, fragmented, whatever you want to call it, the old certainties – if ever there were – are long gone.

The world of work is no less confusing. Job for life? Forget it. These days you can spend years of your life giving your all to an organisation only to find yourself surplus to requirements one random Friday morning. We seem to be working harder and longer, maybe even earning more money, but the trade-off between life and cash is beginning to seem like a rather raw deal. These days, we're told that we are the engineers of our own careers, in charge of building, maintaining, training, negotiating, securing, styling and promoting the brand called 'me'. But none of our bosses ever gave us a break on budget targets or project deadlines because we needed time for 'me'.

Then there are the kids – every life experience, from birthdays to rainy Sundays, the target of advertisers and marketers. They're being raised in a world with so many choices – from the Wiggles through to Bob the Builder, from PlayStation® through to Xbox – but so few meaningful ones; their days are crowded with manufactured experiences and brand-extension opportunities. If we adults are confused, how are children supposed to learn positive human values, judge what's important and what's not, and learn the skills that are going to be needed to navigate this ever more demanding environment?

It's enough to make you wonder what is happening in our own great game of *Civilization*.

And one more thing ...

Late one night, after a few beers, Mike called Adrian – just starting his day at the office in London – and assaulted him with a stream of consciousness that sounded much like the paragraphs above. Understandably, for someone trying to actually work, Adrian was a bit impatient with it all: 'Oh, what a load of moaning old bollocks', he said. 'We've never had it so good.'

'We're not starving, just the opposite – we're getting fatter every year. More of us than ever live in comfort, with all mod cons – if the microwave breaks, we buy a new one. The garage has an electric door, a couple of cars and a refrigerator just for beer. At the touch of a button, in the comfort of our home-office, we can get information, entertainment or education all delivered from anywhere in the world at standards that were inconceivable even a few years ago. We can travel on a whim to the world's great beauty spots or sporting events. And if the job you left before you went away is no longer there when you get back, there will probably be another one along in just a minute.'

Adrian continued: 'Life expectancy is swiftly heading towards triple figures – and what a life. Standards of health and welfare are higher than our parents could even imagine. For the fortunate rump of humanity, the golden billion, this century offers medical breakthroughs that could see pain, suffering and illness pushed back into folk memory. What with cosmetic surgery and botox injections, even the physical reminders of our mortality can be removed, or at least disguised.'

'So don't be ridiculous, Mike. To sit there on the phone 15,000 miles away in Australia, surrounded by abundant evidence of our progress and prosperity and grumble at the state of things and say we've got it hard is total BS.'

Yes, perhaps the newspapers today weren't brimful of great news, but they hardly compare to the Great Depression, Hitler's crimes, Stalin's gulags, the Cultural Revolution, Pol Pot's Year Zero. It would be hard to argue that baby-boomers and above, born in the Western world from the mid-1900s onwards, aren't history's most favoured sons and daughters.

It shut Mike up at least. But over the next few days, even the ebullient Adrian admitted that in his darker moments he gets that niggling feeling of foreboding, a dread that the game may be up and that we're changing things for the worse. This, we decided, is a crucial aspect of what we came to call Crunch Time – and it's all the more threatening for the fact that our everyday lives, carried out in a world of abundance and impulse, have done so little to prepare us for its challenges and because the nature of the change is so unknown.

A few more calls and emails later, we decided that there are good arguments for saying that the near future is going to be the most interesting ride the roller coaster of human history has ever taken. It will be interesting not just because it will be turbulent – when has human history been anything else? – but because there's a definite sense that as a species we're facing collective decisions that will prove critical to our future on this planet.

And, more to the point, even if you don't believe that

the 21st century is more significant or important than any other period in history, one thing undeniably sets it apart from other times: these times are *our* times, and it's *our* collective actions that will shape future history. It's *our* news that appears in the papers and on television every day – so if we're living through the next great extinction, then it's *our* problem and *our* children's problem; if we're moving into a post-human future, it's our humanity that will be transcended or rendered obsolete; if the world is overheating, it's our children and our children's children who will face the climate's changes. That's why we need to explore and understand these things, so we can make the choices that will give us the greatest chance of passing something worthwhile on to generations that follow.

Crunch Time is an investigation of *our* period of history, the age upon which our ideas and actions will stamp their mark. It's based on a radical idea – that our period of history is special, that we're going through a transition that's somehow important. The problem is that surely everyone who ever lived felt this way. But all around us there are signs that it *really* is different this time. In the first chapter – 'Why is it Crunch Time Now?' – we offer four objective reasons why things are different.

The early 21st century does not feel like a period of great optimism, vibrant cultural renewal, renaissance or rebirth. It feels like the best we can hope for is more of the same, and the worst we will face will be much, much more dire.

Despite the wonders of our technology and the enormous leaps and bounds we have made in economic and social progress, it's almost as if the bad guys have won. Millionaire hippie John Lennon was wrong: we need a lot more than just love – we need low interest rates, cheap energy, missile defence systems, pre-emptive strikes on terrorist training camps, and detention camps for asylum seekers.

Meanwhile, our refrigerator of beer remains full. Those of us lucky enough to be born into the Western middle classes have all our basic physical needs fulfilled: we have food, shelter, freedom from need, really, and now we ought to be turning our attention to more worthwhile stuff – building better futures for ourselves, our families and our friends. In shrink-speak, the bottom of Abraham Maslow's pyramid of needs is built and ready to hold all the better stuff that's supposed to come on top. But for some reason that stuff seems ever more distant, drowned out by the MP3 ring tones of our new phones, hidden beneath the ever-proliferating supplements of the Sunday papers that few of us actually read.

In a world where you can satellite navigate your way to the fridge, we have the paradox of an ever more sophisticated lifestyle that's consuming itself to the point of collapse. This book is an investigation of that paradox.

How to read this book

What we aim to do is look in turn at the big issues of the decades to come. Taken together we can see that they have common threads which give us an indication of where we are headed, the forces that are pushing us that way and, most importantly, why. Conclusions are difficult to come by exactly because certainty in this day and age is difficult to come by, but we do get there. This book is not a polemic, not a partisan argument for a particular solution to a particular problem, because – let's face it – partisan beliefs, ranting polemics and recipes for solutions are *so* last millennium.

That doesn't mean that we're left completely floating and rudderless. Our answer to the confusion that reigns in the big picture is to disengage from the big ideas and look closer to home, to our own lives and those of others who have been through equally confusing situations and come out the other side with something to show for it. In the chapters that follow, we look at the big picture but try and bring it all, in the end, down to the local, to the things that we can change within our own lives. It means focusing on human characteristics that the 20th century all too often forgot.

In 1966, in front of a crowd of students at Cape Town University in South Africa, Bobby Kennedy spoke about the conditions that compel attention and action in every age, in every country and in every generation: oppression, social injustice, violence, danger and uncertainty. He was speaking out against apartheid in a country that wouldn't repeal its racial laws for another thirty years. Nonetheless, his message was a simple one, and this book shares it: 'Few will have the greatness to bend history', he said. 'But each of us can work to change a small portion of events, and in the total of all those acts will be written the history of this generation.'

Dr Seuss said the same thing in *The Lorax*, but he said it shorter:

> Unless someone like you cares a whole awful lot,
> Nothing is going to get better, it's not.

CHAPTER 1

Why is it Crunch Time Now?

Speed, contagion, risk and fragility

Few will have the greatness to bend history, but each of us can work to change a small portion of events, and in the total of all those acts will be written the history of this generation.

Robert F. Kennedy, 1966

In that same speech, Bobby Kennedy popularised an ancient piece of eastern wisdom. Three years after his brother had been shot dead and a couple of years before he too suffered the same fate, Kennedy told his audience: 'There is a Chinese curse which says, "May he live in interesting times". Like it or not, we live in interesting times ...'

Kennedy's short life may have been true to the curse, but the curse itself is based on a lie. Any Mandarin speaker will confirm that there's no such curse in Chinese. It's a 20th-century invention of an American political speechwriter. Not Chinese, not ancient, but nonetheless resonating with an ironic wisdom that sounds like it's been around forever. It's a great metaphor for understanding Crunch Time – if ever there were interesting times, these are they.

We live in interesting times, but so what – hasn't everybody from the dawn of history? Why is the 21st century any more serious, critical or even interesting than those that have come before?

We offer four reasons why now is different, why the 21st century will suffer from the Chinese curse more than any that preceded it. Today the world is faster, more contagious, riskier and more fragile than ever. These are the fundamental factors that make our time the most critical yet.

Stop the world, I want to get off

Remember Archbishop Wulfstan a thousand years ago, admonishing his Viking-ravaged flock with the gloomy words: 'The world is in a rush, and is getting closer to its end.' How would he feel now? Crunch Time accelerates every day. Change and innovation happen faster, knowledge develops and obsolesces faster, decisions are made faster, and their consequences – intended and unintended – become clearer in a shorter time span. And those consequences shape and surround our lives.

They influence the way we live, the values we hold dear and the relationships we form. They matter to the permanence and credibility of the institutions around us that provide support, work and structure, including families, employers and governments.

You don't have to be a genius to see it. Over the course of the last hundred years, the developed world has seen improvements in transport (from horse and cart and steam train to mass jet travel and even space tourism), communications (from the telegraph to wireless radio to satellite television) and knowledge (from basic literacy as the exception to mass higher education) – all of which have simply made the world a faster place to live in.

On 17 December 1903, the Wright brothers made the

first powered aeroplane flight at Kitty Hawk. It lasted twelve seconds and took them 120 feet. By 1919, Australia was offering a massive cash prize to anyone who could fly from Great Britain to the great southern continent in less than 30 days. Half a dozen people died trying. Now the journey takes less than 24 hours, and a couple of thousand of us make it every day, with deep vein thrombosis the only statistically significant cause for sweaty palms. Some say the economic optimism of air travel is in retreat. When the last Concorde was abandoned in November 2003, the world stopped shrinking and started expanding once again. The airliners of the future won't fly on ethanol or nuclear power.

Still, communications fill the gap. In the middle of the 19th century, a jobbing portrait painter from Massachusetts tapped out 'What hath God wrought?' in the first ever telegraph message. The code that Samuel Morse invented carried a thin stream of messages to bring instant communication over long distances. That code officially disappeared a few months before this century opened. Today, in place of a series of dots and dashes, we have ones and zeros digitally relaying simulcasts of rock concerts, sports events, news conferences and terrorist attacks watched by millions, even billions, around the world. This is made possible by communications satellites orbiting the earth. The first of them began circling the globe in 1969. Today hundreds of them criss-cross the heavens, littering the skies.

And as for knowledge, we're told that at the beginning of the 21st century it's the single most important industrial sector in the developed world. In 1900, the overwhelming majority of people in developed countries worked with their hands. They were farm workers, builders, factory hands, miners or servants. By 1950, that figure had dropped to about half in the United States' labour force. Now, less than a quarter of America's workers make their living from any type of manual work. The same is true in the rest of the developed world – more and more people are working in the amorphous world of bits, rather than the more concrete world of atoms. And it matters.

Over the course of the last decade, the addition of one further critical factor has accentuated this trend: communications technology, including email and the internet, has pushed the ride even faster. The network created by a global web of email addresses and servers has revved up the pace of life simply by increasing the amount of communications and information available to each and every one of us. It's a faster world for us, a faster world for everybody. In Chapter 10, 'Extreme Evolution', we see how information technology and communications are driving the exponential growth of scientific knowledge, and explore what it means for us.

These three forces – transport, communications and knowledge – have raised the pace on the global treadmill from a gentle jog through to an exhausting uphill sprint. The world economy isn't only faster; it makes life faster for just about everyone who is engaged with it. And these forces of speed and change make everything contagious, from social movements to social diseases.

Action, infection, contagion

A few months into the 2000s, a small group of truck drivers, hauliers and farmers angered by fuel price increases in Britain began a movement that copied the protests (about something completely different) of French farmers that they'd seen on TV. As other angry motorists in Britain saw the fuel protests on the news, they too joined in. The popular uprising came close to bringing one of the world's biggest economies to a grinding halt.

A few months earlier, an international collection of misfits, with barely a goal in common but a shared antipathy towards the WTO, created such violence and confusion in the streets of Seattle that one of global politics' dullest events became a symbol of the bankruptcy of world governance. Masked, black-clad riot police took to the streets for subsequent meetings and clashes in Prague, Genoa, Melbourne, anywhere in fact, where the world's leading politicians and their retinues met – generating serious debate inside the meeting rooms about the validity of the protesters claims. International travel and international communications made one of the oldest forms of protest – street demonstrations – internationally contagious.

Contagion can arrive at your desktop too. In 2003, a bored web designer from North Wales was sentenced to two years in prison for designing and distributing viruses that hit 27,000 computers in 42 countries. That's small beer compared to the handiwork of 'Mafiaboy'. The teenager from a genteel suburb of Montreal used distributed denial-of-service (DDOS) attacks to bring down the websites of CNN, Yahoo, Amazon, eBay and Dell, among others. Although Mafiaboy's technical skills were limited, at his court hearing experts estimated the young man had been responsible for damage, lost business and costs of over a billion pounds. From his bedroom, all by himself. Quite an achievement.

In the economic arena, this new kind of contagion is well documented. It's a vicious circle that plays itself out repeatedly: in Mexico, Russia, Asia, Argentina. In a world of huge capital, technology and trade flows, it's no surprise that 24-hour stock markets have forged a greater level of interconnection between countries and regions. Yet in times of crisis, those connections just carry chaos more quickly.

Protests, computer bugs and investor confidence may be highly contagious, but diseases are even more catching, diseases we didn't even know existed, like Nipah virus. At the end of the 1990s, Nipah virus killed over a hundred people in Malaysia and led to the slaughter of a million pigs. It was new to people and pigs but not to the flying foxes that were the main carriers: for them it was endemic. Their saliva infected part-eaten fruit; their droppings and urine carried the virus too. In their normal habitat, this wasn't a problem. But as they moved from disappearing jungles to feed in orchards, they came into contact with Malaysia's vast industrial pig farms. The pigs caught the virus, and passed it on to farm workers. The mortality rate ranges from 40 to around 75 per cent and there is currently no known vaccine or cure. Subsequently there have been outbreaks in India and Bangladesh.

In the past decades, scientists have discovered previously unknown infectious diseases for which there is no effective treatment at the rate of one a year. HIV is the most appalling and spectacular example – unheard of before the 1980s, it's now killed millions worldwide. SARS, a respiratory virus, brought South-East Asia to a standstill at the beginning of 2003. Over the next few months, the illness spread to more than two dozen countries in North America, South America, Europe and Asia. It was brought under control by isolating patients and finding a treatment, but it was one of half a dozen outbreaks in the previous five years. Now medical nerves jangle at avian flu in the form of H5N1, or a resurgence of smallpox.

And diseases enjoy the travel benefits of globalisation as much as humans do. The Black Death spread through the medieval world in ships' cargo holds. The Ebola virus skipped from an African village to a German isolation ward in a single plane ride. Passenger planes carried the insects that brought West Nile Fever to the United States.

In early 2002 in the UK, a new government agency was unveiled to manage and combat the spread of infectious diseases. It was needed, said the government's top medical adviser, because diseases have become tougher and tougher to detect and monitor. His reasons included big increases in world travel, new technology-related illnesses, global warming, changing sexual behaviour and new drug-resistant organisms.

So, Crunch Time is a contagious time, and that makes it interesting. But interesting times aren't necessarily bad times. Knowledge, understanding and culture – the finer things in life – cross borders as easily as germs.

At the time of writing, one of Mike's close friends was having a bone marrow transplant to treat his cancer. The bone marrow was flown from Spain to Sydney, the closest match they could find.

The frenzy of creativity that surrounded the dot-com boom of the late 1990s – short-lived and unprofitable as it may have been – demonstrated that the internet was still capable of spreading knowledge, as well as viruses, faster and further than at any other time in history. Despite the investment crash, we now buy our weekly shopping, books and holidays online efficiently and cheaply.

What all of the above examples illustrate is that we're now living in a world where everything from political movements, economic trends, fashion, communications, and disease operate at a higher level of contagion. Th spread faster and farther than ever before. When the United States economy sneezed following 9/11, the rest of the world caught a cold. Why? Because communications, information technology, globalisation and liberalisation have created a world that's much more interconnected. These things spread like forest fires. And as the people of South-East Asia know – the smoke from raging wildfires can choke cities thousands of miles away for months.

The Chinese curse manifests itself other ways too. There are objective reasons to believe that the world we live in is riskier than it's ever been before, for the individual, for organisations and for nation states. What does it mean to be riskier? It means that when things go well, they can go very, very well, very, very quickly. The exponential growth of the internet and the economic boom that followed it showed us all how the modern world can offer a huge upside. But equally, when things go bad, they go rotten – witness the speed of economic collapse in Argentina; the ballooning United States' deficit and the resulting threat to economic well-being all over the world; and the spread of militant Islam and the shadow of terrorism it casts.

Hurricane-force change

The global insurance industry is world capital's warehouse for risk. It's in a crisis that's been growing since the beginning of the 1990s, and the crisis has been caused by natural disasters.

Hurricane Andrew, which hit Florida in summer 1992, was by far the costliest loss the insurance industry had ever seen – US$22 billion at current values. That's $2 billion more than the costs of repairing New Orleans after Hurricane Katrina.

Before Andrew, insurers had assumed worst-case losses from a hurricane to be $8 billion. They came up with that assumption when Hurricane Hugo hit Florida in 1989. Hugo cost a mere $5.4 billion but took out several big insurance companies almost immediately. Hurricane Mireille in Japan in 1991 came in at $6.5 billion. $8 billion just seemed like the next plot on the graph.

Insurance is more than just a convenience – it's the grease that lubricates capitalism. The two aeroplanes that destroyed the Twin Towers on 11 September 2001 didn't just cause massive and tragic loss of life. They also cost insurers $70 billion. In the wake of 9/11, as insurers folded or stopped insuring stuff, all over the Western world, the fabric of daily life was slowly unravelling, as small businesses from retail malls through to tour outlets were forced to close because they couldn't find insurance. Post-9/11 premium rises affected pretty much every business on the face of the planet: in the face of multi-million dollar liability risks, aeroplanes stopped flying, manufacturers stopped making things, people's daily lives were affected in the most fundamental ways.

The world in which we live is now more concentrated and interconnected – in terms of wealth, population and economic and social activity – than ever before. The reason why Hurricane Andrew was such a disaster was because lots and lots of rich people were clustered together in one part of the world – so when Andrew hit, the impact on those people and what they owned was enormous. In the few blocks around the World Trade Center, the same was true. But around the Indian Ocean the reverse holds, and the tsunami that killed over 200,000 people in 2004 barely affected the financial world.

The financial markets, too, are suffering from this new

riskiness. Share prices have become more and more volatile – the riskiness of the world's top four share markets steadily increased over the final twenty years of the last century. And, because of contagion, they move together, so when one plunges, it infects all the others. More than that, not only do stock markets move more closely together these days, they matter more. Worldwide, stock market capitalisation counts for more as a proportion of world economic output than it ever has before. Between March 2000 and March 2001 – the zenith of the dot-com crash – $10 trillion of paper wealth was destroyed. Never has so much been lost in such a short time. But those losses have all been made up for since then, in the boom years that have followed.

Two things drive this increased level of risk, both of them to do with our lifestyle. In the financial world, one is an increasing separation between what's happening in the real world, and what's happening in the fictional world of money. At the same time, as we will see in Chapter 4, 'Envirocide', the risks to our life and health are driven by the fact that we live at the limits of nature – our surrounding material environment is almost entirely the product of human intervention.

So, we're living in a world that's faster, more contagious and extremely volatile. For us, that combination spells trouble. But that's not all.

This side up

Life in a fast, contagious and risky world feels as though there's little any one of us can do to tame it. The world feels more fragile than it's ever been. Everywhere you look, our public institutions – the government, democracy, international organisations – are seen as discredited, inadequate and simply not up to the job.

Along with the advances of the last part of the 20th century, including information, communications, globalisation and liberalisation comes the understanding that human society and its interactions with the world in which we live are simply too complex to be governed effectively.

Certainly, at the global level, those institutions have proven to be incapable of addressing the challenges this new world has thrown at them. From military interventions in Somalia, Rwanda, Bosnia, Afghanistan and Iraq, through to the complexities of global warming, population control and Third World debt, they have all too often been accused of making things worse rather than better.

Until the end of the Cold War, the UN was considered a sterile but essentially harmless institution. Since then it has been sidelined by its key sponsor and host nation, as an inadequate means of implementing American policy. It still churns out study after study containing ever more statistics, but its hand-wringing over refugees, education, health and the environment are only rivalled by the disregard of the world's most powerful states for its opinions and efforts. Its Security Council reflects a balance of power long past, and its inclusive membership gives weight to dictators and democrats. The conclusion of many is that the UN is a hollow talking shop.

The International Monetary Fund (IMF) and the World Bank too have failed to live up to their original promise. Their mandate when they were created in 1944 was to help prevent future conflicts by lending for reconstruction and development, and by smoothing out temporary balance of payments problems. They were to have no control over individual governments' economic decisions nor did their mandate include a licence to intervene in national policy. Both now suffer from indelible and irreparable image problems (read faults). Critics say they are consistently used as a tool to promote Western ideals and military causes; that because they serve the interests of their Western masters, they can have little international legitimacy. But their money still talks.

In the sphere of economics, government after government has come a cropper after genuinely well-meaning attempts to manage boom were followed by bust. But the problem has always been, and always will be, a lack of knowledge – the more we know, the more we know we don't know. Information about what's happening in an economy is approximate and always out of date; and the tools a government possesses to control economic activity are crude and brutish: taxation, spending, interest rates, currency circulation. Each of these takes a long and unpredictable amount of time to produce any effect, and the only genuinely discernible result of many economic management decisions is that – with hindsight – they appear to be exactly what wasn't needed at the time.

Politically, even when it's widely agreed what measures need to be taken to control particular risks or resolve certain issues, it's rare that the actual machinery of government will allow action to be taken. Look at the glacial progress that's been made towards achieving the environmental objectives agreed upon at the Rio Summit in 1992, and the hash-up that was the Johannesburg Summit a decade later. The mismatch between local needs and international understanding and agreement is too wide to be bridged by our current institutions. Yet we're far from even beginning the process of redesigning the fragile governance structures that many would say are leading the world toward disaster.

Everywhere you look, the signs of government fragility are showing through our social fabric. Witness the increase in numbers and power of non-governmental organisations. Look at the sharp and disturbing increase in the number of walled and gated communities – huge numbers of wealthy individuals in the United States and beyond are using their purchasing power to remove themselves from society as a whole – creating a patchwork of suburban mini-states. Look at the resonance militant Islam has with young people born and bred in our own society.

You, me, them, everybody

The most difficult aspect of many of the issues and risks this chapter describes is that their causes are deeply embedded in the way we live – in what it means to be an active member of Western society at the beginning of the 21st century. They can't be isolated and extracted from the everyday actions of everyday people. If the simple act of driving the 4x4 to work is a direct contributor to rising temperatures in someone else's country, searching for solutions means looking beyond the way we make laws, products and lifestyle choices.

Meanwhile, in searching for solutions, even if we knew everything there was to know about one specialisation, that wouldn't be enough. Poverty might be the result of poor economics, but it's also the result of different types of social interactions, cultural issues and physical processes like transport, pollution, education and health. That's why so many of the sciences are at loggerheads with each other – look at the debates over the environment and economics, free trade and globalisation, security and freedom. All of these, and the rest, are the results of serious disagreements between educated people with valid perspectives on a tangled world.

Perhaps it's this insight – that everything is connected and we are connected to everything – that has driven the increasing popularity of systems thinking. The 21st century will see huge growth in the study of complex systems in everything from management theory to cosmology. At the same time, it also explains the persistence of religious beliefs, and in parallel, the rise of New Age holistic thinking.

For the first time in history, we find that the decisions we make as ordinary people actually do make a difference. Cumulatively our everyday lives are writing the history of our species even before we have lived it. It may or may not be the conclusion, but the 21st century is surely a new chapter and an exciting one.

CHAPTER 2

Too Many Toys

The successes and excesses of consumer capitalism

The tyranny of toy cupboards

At the time of writing, between us we have five young children. And they all have too many toys. Huge mounds of unbiodegradable plastic that clutter up the corners of their rooms, fill the Ikea drawers and plastic bins, beeping and whistling when you touch them.

It's not that our kids have particularly indulgent parents, or relatives who work in toy stores. The prima facie evidence we have spied on parental excursions to other toddlers' bedrooms confirms that the toy oversupply at our houses is in no way an exception. Friends' birthday parties, Christmases, rainy weekends, visits from Grandparents or family friends, a trip into town, feeling blue, a scuffed knee, all of these and more are excuses for new toys, toys that cost nothing, provide varying levels of joy and will sooner rather than later end up in landfill. The reason for this is simple – we're richer than we have ever been, and toys are cheaper.

It's basic economics: the richer we get, the cheaper they are, the more toys we buy. The Americans, for instance, although they comprise only 4.5 per cent of the world's population, buy 45 per cent of the global toy production. The average American child gets 70 new toys a year.

Think about that. All over the developed world, from Kent to Kentucky to Kanagawa, there are middle-class children with too many toys. There is a toy glut. We are drowning in beeps and whistles, and it has yet to make the papers. And it doesn't take much imagination to stretch this metaphor from the toys in our kids' rooms to the toys in our own living rooms (HDTV, movie downloads, SDSL), the type of holiday we take (wreck diving in Florida, Club Med Phuket), or the toys parked in the drive (4x4, jet ski, ocean kayak).

Like it or not, the success of capitalism has brought us, in the developed world, much to be grateful for. Between 1950 and 2000, the rich world grew by an average 2.5 per cent a year, a rate that saw people get four times richer than they had been. Today most of us ought to want for practically nothing, really. Barely a home in Britain and most of the rich world is without a refrigerator; a house without a television set is rarer than hens' teeth; most households have a car and almost half have more than one. Just 50 years ago, these percentages would have been reversed.

Our aspirations have come way beyond a car in the garage and a colour TV to mobile phones for the kids, high-speed networked internet connections in every room, surround-sound HDTV, walk-in closets stuffed with throw-away clothes we wear only once. Welcome to progress.

But this scale of economic growth, which feeds directly into the number of toys in our children's bedrooms and the number of cars parked outside, is a new innovation in human civilisation. For millennia – right up until the industrial revolution and the British Empire – whatever economic growth gave, population growth took away, keeping average living standards pretty much constant from the dawn of history.

Since the late 18th century, though, countries have in turn taken on the economic structures of a modern state and taken off. The process began with the original industrial revolution in England, then the rest of Europe. By the end of the 19th century, Germany and the United States had outstripped Britain's manufacturing production and the revolution had reached out as far as Japan. The rest of the world has also sooner or later, and to a greater or lesser extent, joined in the march to modernity – even the famously isolated mountain kingdom of Bhutan is now importing the marvels of satellite television and the stock market.

When countries start on the path to economic modernisation, miraculous things happen. Roads get paved, cars appear to drive on them, electricity is generated, peasants move from the fields to the cities which house and feed them (with variable success), communications networks get installed, illnesses are cured, TV sets pop up in households, children's bedrooms fill up with toys. Whatever the downside for those of us drowning in excess, for those on the *real* downside, economic growth *is* a wondrous thing.

Capitalism case study: Botswana

Take the quirky case of Botswana. Botswana is the size of France, but with barely a couple of million people. It's located within the world's poorest area – sub-Saharan Africa, an economic and general disaster zone.

But it happens to be the country that has seen the

highest sustained rate of economic growth per person *in the world*: in the last third of the 20th century it grew faster than Singapore, the United States, anywhere.

A former colony, Botswana wasn't exactly left primed for growth by the British Empire. In 1966, the year the English were celebrating their only World Cup soccer victory, Botswana was granted independence with one abattoir, a mile and a half of paved road, two schools and just enough university graduates to form a couple of football teams.

Despite these handicaps, the country has got richer at a blistering pace. A lot of that is down to diamonds. They account for 40 per cent of the country's output. And a lot of that money has gone to make the Botswanan rich even richer. Botswana is no egalitarian paradise.

Still, the money hasn't all disappeared into casinos at Monte Carlo and villas on the Côte d'Azure. The average Botswanan has done pretty well. Since the British quit the country, the adult literacy rate has doubled to nearly 70 per cent, primary and secondary education enrolments have rocketed and life expectancy has leapt up by eighteen years. Economic growth has delivered a country where people live longer, more educated lives and the country as a whole is producing many times as much stuff. Travels to Botswana these days reveal a stable and advancing country with educated and happy people. Gaborone, the country's capital, is dotted with smart buildings and new malls. Look at neighbours like Zimbabwe and see the difference economic growth can make. Botswana's problems are far from over. It has the highest rate of HIV infection in the world, with over a third of its population affected. But you have to think that it can face challenges better with a well-educated and literate population.

When the world's wealthiest man, Bill Gates, was looking for the most promising place to put some of his billions of excess funds, he chose Botswana, giving the country over US$50 million in 2003 to fund an anti-retroviral drugs programme. AIDS drug manufacturer Merck, who provided both the drugs and the management of the programme, matched that money. Without the infrastructure and education provided by Botswana's strong growth over the previous years, it's unlikely they would have been so enthusiastic.

For those who wish we'd never crawled out of the caves, Botswana shows that economic growth is a good thing. It's the only force we know of that can bring the billions of poor in the developing world out of miserable existences towards lives less dominated by the privations of poverty with more of the freedoms and opportunities we wish for our own children.

So, growth is good. Hold that thought.

How to get there

'Growth is good' is a bald statement. But it's the foundation upon which most of our societies are built. Politicians get elected and tossed out on their claimed ability or inability to promote growth; media pundits compete to shout the loudest about which policies and politicians will make the country grow faster; countries rank themselves on their growth performance, boasting and chest beating when the numbers are good, or hanging their heads and searching their collective navels when they're not.

And late 20th-century history – with its epic struggle between capitalism and communism – was dominated by the ideological battle about the best way to foster growth.

Most foundation economics courses begin with the

professor standing at the chalkboard and declaring: 'Resources are scarce, and economics is all about how to exploit them in the best way.' Resources are unquestionably scarce – otherwise everybody's kids would have too many toys. Also unquestionably, resources are distributed in an inequitable way. Since economic growth became a feature of developed and developing economies, economics – the 'dismal science' – has focused on how to get around these problems.

To date, two guiding principles have led economists' thinking – and those principles laid the groundwork for the political systems that the West and the East built for themselves. The first is represented by Adam Smith; the second by Karl Marx.

A. Smith v. K. Marx

Adam Smith argued that people enjoy their daily bread not because the baker is a kindly soul with a good heart, but because he's after a profit. According to Smith, society's best interests are served when people are allowed to get on with what they do best, and what will profit them most. Prices provide information about scarcity, and behaviour adjusts accordingly – the invisible hand guides resources to those who can most profitably use them, or those who want them the most and are prepared to work to get them. In the end, as if by magic, social welfare is maximised. This is the foundation of the free-market ideology that guides liberal economies, Britain, the United States, Europe and everyone who wants to be like them.

Of course, there is much the invisible hand doesn't take into account, and the bulk of economic policymaking is concerned with dealing with social problems that the invisible hand doesn't touch. Think of the environment, poverty, the status of women, care for the sick, elderly and unemployed, education and other 'market failures'. But in the main, the message from Adam Smith is that if it ain't broke, don't fix it.

On the other hand, Karl Marx started with the premise that the whole of society is already broke – easy enough to agree with if you look around. Clearly, resources are not distributed evenly to start with, and leaving it that way just makes the difference between rich and poor greater. It takes money to make money, as the old saying goes. Why should the baker benefit from his ability to bake bread, and ownership of the bakery, just because he was born in a certain place at a certain time? Surely, society as a whole should benefit. The more uneven things get, thought Marx, the more social tension will bubble up.

The solution, for Marx, was to place ownership of the bakery and the baker's labour in the hands of the state, and then let the state decide where best these resources should be deployed. The result would be a more efficient use of the resources of the country, and a stronger shot at growth.

It was a nice thought but, as a thousand broken statues strewn across Eastern Europe will confirm, it didn't work. It didn't work because Adam Smith's observation that the baker is innately selfish was right on the mark. Under state control, the baker finds that he can buy the same amount of toys for his children no matter how many loaves of bread he bakes – so why bake a hundred in a day when he might as well bake half a dozen? What's more, the toy-maker draws the same conclusion. So even with his socialised wages burning a hole in his pocket, the baker can't find toys to buy. That's why queuing was considered a core competence in the old communist Eastern bloc – there just weren't enough goods to go around at any price, so it was first come first served.

The 20th century settled this debate firmly on the side of Adam Smith and capitalism. Success in terms of economic power has flowed to countries that adopted the principles of economic freedom. In other words, letting people get on with whatever it is they want to do with the resources available to them – within a framework of basic law and order. Some people say that Smith's victory has been so astounding that the fundamental economic problem of scarcity has been solved.

But this is clearly not the end of the story. Many of the problems we're faced with as we stare down the barrel of the future are the direct result of capitalism's extraordinary success, and the 'market failures' that come with it. Some might say that capitalism has been too successful for its own good.

Why is capitalism eating itself?

Too successful for its own good? If you can never be too rich (never mind too thin), how can capitalism go wrong? There are two causes. The environmental cost will be dealt with more thoroughly in Chapter 4, when we look at the challenges humanity faces preserving the environment that sustains it. Fundamentally, this comes back to the question of limits – if we live in a world that's limited in its size (and clearly we do), then if we keep growing and growing, it's inevitable that we will eventually use it up.

Gandhi perhaps said it best. When India was on the brink of independence, a journalist asked Gandhi whether India would now follow the British pattern of development. Gandhi replied: 'God forbid that India should ever take to industrialism after the manner of the West ... It took Britain half the resources of the planet to achieve this prosperity. How many planets will a country like India require?'

The second issue facing 21st-century capitalism is equally serious, and it's a structural fault within the system itself. By necessity capitalism measures and rewards success – for countries, communities and people – using only one yardstick: money. However, true success is more complex than that.

At least that's what your authors learnt at the London Business School (LBS), where we met.

The Hogwarts of capitalism

LBS is an august establishment located in London's beautiful Regents Park, and is dedicated to one thing – helping its students get rich. Paradoxically, it does this by charging them tens of thousands of pounds in fees (in our own case close to £50,000) to teach them how to squeeze money out of others. It's a simple proposition for potential students: learn how to screw others by being screwed yourself.

LBS is a kind of Hogwarts of finance. At the end of the 1990s when we were there, the enchantment was provided by the longest and strangest economic boom the modern world had ever seen. Money was being magically made everywhere and by everyone (except by people dumb enough to spend that time studying instead of floating an IPO). And it was being made in unusual and unprecedented ways: the dot-com boom was busy installing 27-year-old nerds in priceless mansions on the back of their ability to programme internet code. The professors –

wizards of the dark financial arts – were struggling to explain the phenomena.

At Hogwarts LBS, Professor Gary Hamel played Dumbledore. No long white beard, but a rakish Errol Flynn moustache and wire-frame spectacles. Hamel was the epitome of the intercontinental business-school hero: living in sunny California, teaching in rainy London, hitting the carbon emissions trail to preach in the world's financial capitals in the biggest boardrooms of the world's biggest businesses.

Hamel had an explanation for the magic of business success in the 1990s. He spelled it out in *Leading the Revolution*, a colourful and breathlessly written business tome published at the turn of the century. One company embodied the principles that turned business lead into gold. Not widely known outside corporate circles, it was called Enron.

'At Enron, failure – even of the type that ends up on the front page of the *Wall Street Journal* – doesn't necessarily sink a career ... Controls form the cauldron in which Enron's innovative energies circulate. The heat comes from Enron's ambition ... and from the chance individual dealmakers have for personal wealth accumulation.' That was Gary Hamel in 2000.

Oops. Enron turned out to be the perfect example of capitalism's extraordinary single-mindedness. Less than a year after Hamel's book was published, Enron went belly up to the tune of over $100 billion – the world's largest ever bankruptcy. Enron claimed the retirement savings of tens of thousands of employees and the life of the company's vice-chairman (who blew his brains out). The company's boss, Ken Lay, died of a heart attack a couple of months before being sentenced to 20–30 years of jail for ten counts of fraud. Amazingly, despite a heap of connections with the Bush administration (and a fair few with Clinton's and Bush Snr's), the United States Presidency walked away from Enron untouched.

Enron was just the biggest of a string of companies around the world that collapsed in the years following the turn of the century, taking down with them hundreds of billions of our pension money. WorldCom and Tyco followed Enron in America; in Europe there was Ahold and Parmalat; SK Global in Asia; in Australia One.Tel and HIH. And the rest. The world's largest audit firm, Arthur Andersen, which was supposed to have been vouching for the credibility of these companies, collapsed as a result of Enron, too. All these companies had achieved financial success by sucking in cash from investors on a good story. Some used sophisticated financing schemes, others engaged in old-fashioned double-dealing.

Gary Hamel was hardly alone in praising Enron. Before its collapse, it was considered daring, unafraid, uniquely positioned to succeed amid the speed, risk and contagion of the Crunch Time world. Six years running, *Fortune* magazine named Enron 'most innovative company', and its Chief Financial Officer, Andrew Fastow won an innovation award from *CFO* magazine (and was later sent down for ten years for accounting innovations of a different kind). No doubt, Enron was an innovative company. It was just that its innovative talent was directed towards creating misleading accounts, persuading investors to hand over their cash, persuading its auditors to verify them, and concealing the real state of the company's finances. As Hamel put it in *Leading the Revolution*: 'The system focuses Enron's most ambitious and creative people on creating new wealth that drives the company's market capitalization ever higher.' It just did it crookedly, that's all.

EPS I love you

The key to understanding the great post-boom accounting scandals of our time is contained in Enron's 2000 annual report. The company, it announced, 'is laser focused on earnings per share'.

Earnings per share (EPS) is the number that tells you how much profit each share of the company makes for investors. According to many of the analysts who decide which companies should get money and which shouldn't, this number, or the growth in this number, is the single most important indicator of how well a company is doing. It's the Holy Grail of Crunch Time business.

Enron's pursuit of EPS growth persuaded its managers to create shadow subsidiary companies into which they could shove the company's rapidly rising debts, while keeping them away from the company's official accounts. A deception, to put it bluntly. As Enron's fictitious EPS kept skyrocketing, greedy banks and investors (only too willing to believe the stories about Enron's magic money machines) threw more and more cash at it. The cheaper the money got, the more businesses Enron could buy, the bigger it became.

Enron's management recognised a crucial factor driving the Crunch Time world: performance and growth are rewarded disproportionately. This is a common conception – as we shall see next in 'Sharing the Spoils', it's a fact that our national leaders tend to rely upon a lot. If a corporation, or for that matter a country or a person, appears to be doing well, investors plunge money in, pushing expectations higher than future profits can deliver. When there's bad news, investors quickly bail – the proverbial rats quitting a sinking ship.

It's these problems, inflated by the acceleration, contagion and risk of our economic system, that raise critical questions about the way our economies operate, and whether they are capable of carrying us forward into the future.

The managers of Enron may have been lying about financial performance, but they were doing so for the benefit of their shareholders. One look at Enron's share price performance between the end of 1997 and the beginning of 2001 shows that there were many who did indeed get rich on lies – provided they got out in time.

Unfortunately for Gary Hamel (and perhaps Adrian Monck and Mike Hanley), once you've gone to press there's no getting out in time. As Enron slipped spectacularly beneath the financial waves, Hamel, the evangelist for free markets, became a contrite capitalist. His column in *Fortune* magazine began to reflect concern about a 'starkly secular and ravenously materialistic' society that had 'lost its spiritual capital'.

Perfection of means

For Mike, the story of Enron and the wave of corporate collapses that washed across the globe, and Gary Hamel's reaction to it, was proof that late model capitalism and the way that growth, wealth, money and greed work together are out of whack.

'The focus on money and growth is all very well as far as it goes, but look at what happens when the growth disappears', he said. 'Suddenly we have to find a new foundation for our lives. We find ourselves frantically searching for a new source of "spiritual capital".'

Growth is good; we agreed already – without it, our children are toyless. Nevertheless, the way that we achieve it can be violent and hostile. More than that, a single-minded obsession with growth is as likely to be as ineffective as ignoring it completely. Economic growth isn't the only kind of growth we experience in our lives, and isn't the only kind of growth that's good. In the boom-time of the 1960s, Bobby Kennedy reminded economists that the numbers do 'not include the beauty of our poetry or the strength of our marriages, the intelligence of our public debate or the integrity of our public officials. [They] measure neither our wit nor our courage; neither our wisdom nor our learning; neither our compassion nor our devotion to our country; [they] measure everything, in short, except that which makes life worthwhile.'

Kennedy wouldn't have been surprised to see that measuring broader indicators of how we're doing give less impressive results than simple gross domestic product (GDP) figures.

In response, Adrian replied that the problem isn't so much with the system itself – the battle between Smith and Marx settled that one – but with the way people apply it to their own lives. Wily old physicist Albert Einstein put it this way: 'Perfection of means and confusion of ends seems to characterise our age.' The end is happiness, the means is money but the two end up fused in our minds. We make more money. We consume more stuff. We buy more of everything. Still we feel unfulfilled.

In reality, simply making more money, certainly on a national level, doesn't make us happier. There's plenty of evidence that shows that after we've achieved a certain level of subsistence – £10,000 a year per head, according to the statisticians – further increases in GDP per head make less and less difference.

The evidence from surveys shows that rising incomes does everything but make people more satisfied with their lot. On average, people in America, Europe and Japan are no more pleased with their lot than in the 1950s. In Britain, for instance, people are now almost three times better off materially than they were half a century ago, but overall well-being has hardly budged – in fact, 60 per cent of Britons say they can't afford to buy everything they 'really need'. The same goes for the richest people in the richest country in the world: Juliet Schor in *The Overspent American* reported that a quarter of Americans who earn over $100,000 a year say they can't afford to buy everything they really need. Pardon? In Japan, real GDP per person increased sixfold between 1958 and 1991, but the Japanese remained a miserable lot.

In fact, the correlation between money and happiness is very weak, or even negative. In Asia, for instance, the richest countries, such as Japan and Taiwan, have the most miserable people, while the Filipinos – without so much as a pot to, uh, spit in, consistently report themselves happier than any other culture.

So getting richer doesn't automatically make you happier. Why not? One explanation is that people quickly get used to their new conditions, so once you've got that inside toilet or that air conditioning or that DVD player, after a brief flush of happiness the satisfaction wears off. Now all these things that were once considered luxuries are considered essentials.

The other reason that people in richer countries aren't necessarily happier is because people's happiness doesn't depend on their own incomes, but actually on their incomes in relation to other people. Surveys routinely find that people would rather earn £30,000 in a society where £20,000 is the norm than £40,000 in one where £50,000 is the average. You may be over the moon when your boss tells you about your bonus, until you find out that Bruce got a bigger one, at which point you get cross.

It seems that keeping up with the Joneses is an all-important human driving force. Add to this our fear of losing what we've already got, and it's easy to see how we've created our own little prison of materialism. Like the punishment of Sisyphus, doomed to push a rock up a mountain for eternity, we find ourselves frantically re-defining standards of success, achieving them, and then starting all over again.

The Crunch Time search for happiness

> It's time we admitted that there's more to life than money, and it's time we focused not just on GDP, but on GWB – general well-being. Well-being can't be measured by money or traded in markets. It's about the beauty of our surroundings, the quality of our culture and, above all, the strength of our relationships.

These are not the ramblings of a hippie, a church leader or a lefty academic, but the leader of Britain's Tories at a conference hosted by Google in May 2005. David Cameron may be the highest-profile proponent of the happiness agenda, but expect to hear more of this kind of sentiment from politicians and business leaders. Across the rich world, the idea that we should be searching for happiness rather than simply wealth is moving to the centre of public debate, not only in politics but in areas as diverse as jurisprudence, organisational theory, medicine and economics.

In the US, political analysts say the dominance of the Republicans is down to the 'happiness gap', with some 45 per cent of Republicans saying they are 'very happy' against only 30 per cent of Democrats. In Australia, the Labor party is under fire from critics as well as from within its own ranks. Commentators such as the Australia Institute executive director Clive Hamilton and Labor's finance spokesman Lindsay Tanner have targeted the party for focusing too much on the politics of class war and deprivation rather than on the new challenges of affluence.

Amazon.com lists more than 4,000 books with the word 'happiness' in the title. A rash of reality TV shows, how-tos and documentaries have tackled the subject – how people are pursuing happiness, what it is, how to get it. And the concept is influencing the way companies approach their customers (in their marketing, their 'customer relationship management', and their product and service offerings) and treat their employees.

In short, happiness has become an industry, with its own dynamic of production and consumption, its own vocabulary, and its own performance management criteria.

Psychologists and philosophers reckon happiness matters because it lies at the core of everything we do – we're hardwired to seek happiness. According to Harvard psychologist Daniel Gilbert: 'Everyone who has observed human behaviour for more than 30 continuous seconds seems to have noticed people are strongly, perhaps even primarily, perhaps even single-mindedly, motivated to feel happy. If there has been a group of human beings who prefer despair to delight, frustration to satisfaction and pain to pleasure, they must be very good at hiding because no one has ever seen them.'

Happiness is important because it's good for us. Things that make us feel good – food, sex, friendship, winning – are good for our survival, while those that make us feel bad – pain, hunger, illness, losing – help bring us to a premature end. Philosophers and psychologists have put happiness at the core of their theories of human behaviour from the beginning of time. Plato, Aristotle, Thomas Hobbes, John Stuart Mill, Jeremy Bentham, Sigmund Freud, the Epicureans, the Stoics, the Hindus, Buddhists, Christians, Jews, Rastafarians and Zoroastrians have all put it at the centre of their world view, because if they hadn't, no one would have listened to them. Leading a 'good' life was sold as the route to happiness, in this life or the next.

Many studies have confirmed that being happy is good for our health and longevity. Happy people fall ill less often than unhappy ones, and have permanently lower levels of stress. They live longer. Actors who win an Academy Award are on average likely to live for four more years than those who are nominated but don't win. And in one study of nuns that began in 1932, those who were naturally cheerful lived significantly longer than those who were by nature grumpy.

Bring the subject of happiness up at a dinner party and you will discover a fundamental issue: everybody has a different definition. For social scientists, part of the problem is that philosophers and theologians bring value judgements to the process of defining happiness.

John Stuart Mill said that it's better to be Socrates dissatisfied than a pig satisfied, and if the pig is of a different opinion it's only because he doesn't know better. But economists and psychologists beg to differ. For these social scientists, happiness is happiness. It is, according to Richard Layard, an economist at the London School of

Economics (LSE), 'feeling good – enjoying life and wanting the feeling to be maintained'. And happiness is just as valid, for the purposes of analysis, if it's caused by a nose full of cocaine, lifting the World Cup, or helping a senior citizen across the road.

This type of thinking claims that there may be different types of happiness but they can all be compared and they can all be revealed by a very simple technique: asking people.

Professor Ed Diener, a psychologist at the University of Illinois, says: 'It may sound silly, but we ask people, "How happy are you? From 1 to 10." An honest real-time report is an imperfect approximation of a subjective experience, but it's the only game in town ... We can be confident that if we ask enough people the same question, the average answer will be a roughly accurate index of the average experience.'

Advances in neuroscience have recently enabled us to observe that when you say you are happy, your brain is lighting up in the happy places: when you are feeling good, there is brain activity in the brain's left side, behind the forehead.

Scientists are coming up with some concrete conclusions in an area that until now has been nebulous. 'There are two areas in which this research is affecting the real world already', says Professor Andrew Oswald, a happiness economist at the University of Warwick. 'The first is in the courts, where legal systems are adopting economics of happiness methods for valuing the bad things that happen to us.' For instance, we now know that the death of a spouse is 'worth' about US$300,000 a year in compensation, he says.

'The second area is in public policy', says Oswald. 'All

politics is about deciding the right place to spend money – should we spend it on green areas, tax cuts or more facilities? This new work gives us ways of calculating these values explicitly.'

Happiness work can also help businesses understand what will make their workforces better off. We now know, for instance, that commuting is a major source of misery. Andrew Oswald took his own advice and moved closer to work to cut his commute. We know that people simply feel happier when they feel their work has a broader meaning and that they have some control over their destiny. Happiness research isn't solely responsible for these conclusions, but it certainly buttresses them.

Happiness surveys have generated a broad spectrum of ideas and a flourishing genre of literature across almost all the social sciences. The economists have led the way with research showing that although the rich world has been getting richer over the last 30 years, there hasn't been a corresponding increase in net happiness.

LSE's Layard, for instance, in his book *Happiness: Lessons from a New Science 2005*, uses this result to argue for a more progressive tax regime to encourage people to stop working so hard as they grow richer. The same argument is made by the Australia Institute's Clive Hamilton in his books *Growth Fetish* (2003) and *Affluenza* (2005). Both thinkers make a pitch for less economic rationalism and more thoughtfulness in economic and social policy.

Our semi-criminal, semi-pathological propensities

Early economists never considered that we would get caught up in this mouse wheel of production and consumption. Dismal science legend John Maynard Keynes wrote *Economic Possibilities For Our Grandchildren* (1930) to brighten up the dull days of the Great Depression. He forecast that mankind was well on its way to solving the fundamental economic question and predicted that three decades into the 21st century we would be eight times better off economically than in the gloomy 1930s. The long struggle to produce enough to meet basic needs would be over.

Well, we got eight times better off some time ago.

Keynes also predicted that economic success would mean we'd no longer have to slog our guts out to make a buck. Once the economic problem was solved, the great economist ventured, people would work for only about fifteen hours or so a week. A few might work harder in pursuit of wealth, but most wouldn't, seeing the love of money as 'one of those semi-criminal, semi-pathological propensities'.

But what a semi-criminal, semi-pathological bunch we are. Even part-time workers struggle to do only fifteen hours a week. The richer we are, the longer we work, it seems. As we have become richer over the past decades, so we've found ourselves working longer and longer hours. Like a frog in a pot that doesn't notice how hot it is until the water boils, we find ourselves at breaking point. There's something extremely bizarre about this cycle – demand more, work harder, achieve more, become dissatisfied, demand more, work harder. If insanity is doing the same thing repeatedly and expecting different results, then clearly we're all mad.

It may be human nature to be dissatisfied, but – like violence – it doesn't mean we have to accept it or celebrate it. And things seem to be changing. More and more people are waking up to the futility of the rat race (even if you win, they say, you are still a rat) and dropping out.

In Mike's country, a survey by the Australia Institute found that a quarter of 30- to 60-year-olds had decided at some time in their lives that they'd rather earn less and change their lifestyle than stay on the consumer treadmill. The man who did the survey said downshifters 'expressed a desire to do something more meaningful with their lives, and to achieve this aim they considered it necessary to consume less, work less and slow down'. Similar results are showing up all over the world: in a survey in the United States, a fifth of adults said that in the previous five years they had voluntarily decided to make less money and in the UK, polls show similar trends.

Still, our increasing tendency to tune in and drop out notwithstanding, we're working harder, even as we have less and less real need to work. In the developed world, where we have provided for our basic needs many times over, it seems we get more meaning from working than just the money we get in exchange for it – otherwise why would we work so damned hard? But herein lies the rub – if we work for the sake of it, why aren't we working towards something more valuable and lasting than simply filling our bank accounts and buying more toys? Perhaps that's the answer: there's nothing wrong with being devoted to your work ... but if you are going to dedicate your life to something, perhaps it should be something more significant than improving the bottom line for some faceless set of corporate investors, or achieving one target after another. Perhaps it ought to be something that you will be able to sit back and think about with a satisfied smile when you're an old person reminiscing in your rocking chair, or something that you could tell your friends about with pride rather than embarrassment.

Viva la revolución!

At the end of the 1960s, Kenneth Clark, art critic and father of womanising Tory politician Alan Clark, presented a renowned television series *Civilisation* – spelt with an 's'. If he had been a computer game designer he would have called it *Kenneth Clark's Civilization*, but that would have been vulgar and Clark was an aristocrat. With the hubris of an English lord, he reviewed the entirety of humanity's artistic output, and labelled our current state of civilised development as 'heroic materialism'. At the end of the series, having taken viewers from the cave paintings of Lascaux to Guernica, he lamented the emptiness of our culture of consumption:

> It is lack of confidence, more than anything else, that kills a civilisation. We can destroy ourselves by cynicism and disillusion, just as effectively as by bombs ... The moral and intellectual failure of Marxism has left us with no alternative to heroic materialism, and that isn't enough. One may be optimistic, but one can't exactly be joyful at the prospect before us.

As we said in our introduction, dissatisfaction with the point at which we've arrived isn't exclusive to Crunch Time.

Yet even more than when Clark was writing all those years ago, it's all too easy when faced with the complexity of the 21st-century world to feel like the helpless victim, the dupe, the battler against a world filled with injustice and wrongs. But getting past the everyday stresses of just getting to work and sorting out life and actually recognising this feeling is the first step on the path to wielding what power we do have to change things. That change might be evolutionary, incremental even.

We *are* a part of 'the system', we have jobs and responsibilities and everyday worries. But it's membership in good standing of that system that gives us the opportunity to shape our bigger goals: where our pension money is invested, what our societies devote their time and effort to, the endeavours to which *we*, as card-carrying capitalists, wish to lend our support.

In the lucky Western world, we're rich enough not to have to compromise. We understand that we no longer work to put food on the table or fulfil basic needs. Perhaps our heroic materialism is an epic canvas on which to display psychological tropes that were once portrayed on cave walls. Life's about more than being able to buy a warehouse full of toys for our kids; it's about giving them something worthwhile and meaningful to weave into their own lives. That something is knowledge, awareness, values, and a world that is still habitable.

CHAPTER 3

Sharing the Spoils

The limits of globalisation

I dislike feeling at home when I am abroad.

George Bernard Shaw

The goal of the global economy is that all countries should be homogenized. When global hotel chains advertise to tourists that all their rooms in every city of the world are identical, they don't mention that the cities are becoming identical too: cars, noise, smog, corporate high-rises, violence, fast food, McDonalds, Nikes, Levis, Barbie Dolls, American TV and film. What's the point of leaving home?

From a two-page ad placed in the *New York Times* just before the Seattle meeting of the World Trade Organization, which attracted the first and biggest of the anti-globalisation protests

'The unexpected upside of globalisation – you won't have to fly because everywhere is the same', Adrian wrote to Mike about the ad.

We figured there are two ways of looking at this. And the rest ... but let's start with two:

1. The first view says that the reason luxury apartments in Mumbai and Miami are all but identical is because, when presented with a choice between a village hut, a shanty town, or a concrete apartment block, people will generally choose the apartment block. When presented with a shiny McDonald's burger or rice and manioc again, McDonald's will win out. Most people reckon they'd rather drive on an ugly spaghetti junction than not drive at all. These are all just the direct results of man's drive to achieve 'better' things. More than that, it's hypocritical and downright immoral for you, a rich person, to enjoy the fruits of our progress – longer, healthier, better-educated, more independent and richer lives – and deny it to the rest of the world, just because you want to have somewhere interesting to go on your holiday.

2. The second perspective is different. This perspective says that people aren't free to make choices, and the choices they are presented with are determined by forces beyond their control. The choice isn't simply McDonald's versus rice and manioc, it's one type of society versus another, and often choices that are made are done so while staring down the barrel of a rifle – be it real or metaphorical.

The answer of course is that both of these hold true. Where globalisation is concerned, the key to managing our way out of the Crunch Time bottleneck will be to balance these two perspectives, and it's not too much to say that it will be one of the greatest challenges human civilisation will have to face as we head into the 21st century. To illustrate the first perspective, we go to Japan, the second, to Mauritania.

The land of the rising bun

McDonald's has long been the number one target for anti-globalisation's ire. According to lore, it marches across the globe carnivorously swallowing cultures whole, laying the forests of the Amazon waste to feed the methane-producing cows it needs for its burgers, destroying landscapes with its hated golden arches and taking the profits back to America. But McDonald's first foray outside the shores of the United States was not another stop on the long march of Yankee economic imperialism. It was by invitation. In 1971, Den Fujita, a maverick law student out of Tokyo University, took himself to the Golden Arches headquarters in Oak Brook, Illinois and asked for the right to franchise in Japan. Fujita took the brand and localised it – sort of. Ever had a McTeriyaki burger? Delicious.

Fujita was the first person to single-handedly convert an entire culture to a new way of eating. From one store in Tokyo's Ginza district, Fujita's business opened nearly 4,000 franchises, in the process making him one of the richest men in the country. When the company floated on the Tokyo stock exchange in 2001, its prospectus boasted of 10,000 stores to be opened by 2010. All of this, not because of a vindictive orange-haired clown on a crazed power trip but because of the energy and ambition of Den Fujita and the Japanese public's enthusiasm for the whole standardised hamburger thing.

The flipside to this type of globalisation is the countless sushi, Thai, Indian and Chinese restaurants that are on the corner of every street in the rich world. People don't complain about globalisation when it enables two-way cultural flows. The problem is when it doesn't. And it often doesn't.

Make that a quarterpounder with camel cheese

Mauritania doesn't get too much of globalisation's spoils. This poor country is basically the Sahara with added sand, dunes and rock. Some of its poorest people – many living under the World Bank's absolute poverty line of a dollar a day – still live the kind of nomadic existence that seems to reach back beyond bible stories into the prehistoric ages of humanity. Nomadism has a romantic appeal that evokes camel trains, blanket bags, lavishly appointed tents and the kinds of stores that sell ethnic rugs and scented candles. And it *is* romantic, if you're in love with camels.

Mauritania's nomads are camel herders. They live on them and off them. Camel hair is woven into cloth for making clothes, tents and rugs. The dung is burnt for fuel. The hides are cured into tough leather for water pouches and shoes. And camel steaks are a staple of the diet of West Africans.

But, according to Nancy Abeiderrahmane, the owner of Mauritania's first camel dairy, the best thing about a camel is the milk. She says it's naturally low fat and low cholesterol, has as much protein as cows' milk and fewer people are allergic to it. It has a high mineral content and a lot of vitamin C, and is easy to digest because it doesn't curdle. It has half the fat of cow's milk, and less sugar, and is good for diabetics because one of the proteins in it is similar to human insulin.

For the nomads of Mauritania, camel's milk is truly a boon. But the reality of nomadism is that when what passes for camel pasture fails, they have been forced to resort to raiding and slave trading to make ends meet. And if that isn't possible they starve to death in one of those quiet, unreported famines hidden in the statistics, and still of little interest to the news pages and channels of our globalised media. Abeiderrahmane (her married name), was a Brit who arrived in Mauritania in 1970, a year before Den Fujita trekked off to McD's. While Fujita was bringing burgers into the land of sushi, Abeiderrahmane came to see camel's milk as a marvellous opportunity for the Mauritanians. She came up with the idea of making stuff from camel's milk and exporting it, generating hard currency for the hard-up nomadic tribes.

She took her life savings and built Africa's first camel dairy to pasteurise and market the stuff. By the late 1980s, her company, Tiviski, had gone from packaging camel's milk to products including camel's yoghurt and crème fraîche.

Every day, trucks fanned out from three collecting centres in towns along the Senegal River to pick up the white stuff, milked by hand by the nomads scattered around the region. Problem was that because camel milk doesn't curdle, it doesn't easily convert to cheese, so any excess milk can't be stored and goes to waste.

Abeiderrahmane's challenge was how to stop the waste. After attracting the interest and funding of a wealthy Western bureaucracy, the Food and Agriculture Organization, she went to the people who know cheese – the French. She hooked up with a professor from the wonderfully titled Ecole Nationale Supérieure d'Agronomie et des Industries Alimentaires, and together they developed a way to curdle the milk and turn it into cheese, even in Mauritania's brutal climate. The cheese itself has the texture of brie but the taste of goat's cheese. Abeiderrahmane, a good businesswoman, gave the product a romantic nomadic name – Caravane – packaged it up and hopped on a plane and went to Europe with it.

Smart stores there loved the stuff. Abeiderrahmane came away with orders from the likes of Harrods in London and Fauchon in Paris.

Hold the cheese ...

So here was a business proposition aimed at regularising the lives of some of the world's poorest people and developing a trade that could lift them out of the most abject poverty. A European woman had provided the seed capital and ideas, a French professor the technology, an NGO the development funding. It was a model development project. What could go wrong?

Problem: the free traders in Brussels had a thing or two to say about it. The first hurdle was that there was no classification for it. Camel cheese wasn't a dairy product because European Union law says that 'dairy products' only come from cows, buffalo, sheep or goats. Even when this was changed, the bureaucrats decided that they would slap the European Union's highest level of tariffs on the product – despite the fact that there are no indigenous European camel-cheese makers to be threatened. Tiviski was told that for a couple of million euros they could appeal the decision.

Even if the company won on tariffs, because the camels are not mechanically milked, their cheese does not meet European Union veterinary standards designed to keep out foot and mouth disease (of which camels are not known carriers) – so it couldn't bring camel cheese to Europe in any case.

So Abeiderrahmane began looking at other, more distant, markets – the United States and Japan. Again, camel cheese was welcomed in delis and up-scale food stores. But there are no flights there from Mauritania's Nouakchott Airport. Air Mauritanie flies to only one potential market – Paris. And if the cheese transits Paris, whatever its final destination, it's subject to European Union regulations anyhow. So Caravane stays in Mauritania. Nomad income stays pitifully low.

Here it is in spades. Who could benefit more from a more integrated world than the Mauritanian nomads? Better off as camel herders, you say? Perhaps, but tell them that when they lack money to buy sick children medicine, or send them to school. Yet on the other hand we're inviting them to participate in a luxury economy of air-freighted delicacies. Either way, the institutions and infrastructure we have in place in this world of 'globalisation' prevent the Mauritanians from participating, simply because they live on the edge of the world.

From Fujita and Abeiderrahmane, it's plain that globalisation is more complicated than just trade. It's about the spoils of trade, and how they are divided.

Asymmetric trade wars

Trade, reckon economists, is a good thing. According to basic economic theory, countries that specialise in what they are good at and trade with each other will be better off. Take a 'new' country like Australia, for instance. Australia is rich because it digs up and exports its natural resources (like uranium), and grows great swathes of wheat and other agricultural commodities across vast tracts of cheap land. It sends container loads of this to, say, Asia, and gets back cheap clothes and running shoes made in factories where most Australians wouldn't line up to work a shift. Everybody is better off goes the logic, because there are more commodities and more clothes and running shoes than there would be if everyone made their own. And they are cheaper.

The figures back up the theory. In general, in the post-war world, countries that have freed up their economies carefully and selectively have reaped rewards. On the other hand, when countries put up barriers to trade, things go awry.

So far so simple.

Or is it? The post-war world as a whole may have grown in line with the volume of trade, but the rewards of that growth were somewhat unevenly distributed. The reason? Free trade policies in the rich countries were introduced only after their industries were strong enough to take it. In the 19th century, one by one, countries took off on the road to economic modernisation. The common link between these countries was their fierce protection of industries while they were developing and their access to cheap raw materials, in the form of resources and energy.

Victorian Britain is always portrayed as the bastion of laissez-faire, market-led economics. But Britain's proud history of protectionism began as far back as the 14th century, when Edward II brought Flemish weavers into the country and banned the import of manufactured woollen cloth. At the turn of the 18th century, Britain banned wool imports from Ireland and calico from India. It built its own manufacturing base at the expense of destroying these other countries' industries.

Victorian America was hardly different. They slapped tariffs of up to 50 per cent on manufactured imported goods throughout the 1800s, and they stayed high in the first half of the 20th century. Protectionism was a critical contributor to the American Civil War – high tariffs helped the Northern states, but hurt the Confederacy. When the North won, the country hiked import taxes as high as they had ever been. America's professed devotion to free trade – although its protectionist actions still shout much louder than its words – flies in the face of its own economic history.

Even in Asia, where the economic 'miracles' of the 20th century pulled Japan, Taiwan and South Korea out of poverty, it's been because they protected key industries and actively promoted exports. Imports were allowed in only when their own industries had become world leaders.

So the argument that free trade is the only way for countries to develop is clearly bogus. But even if it weren't, politicians tend to be less enthusiastic about trade than economists because they have to be careful about their voters' jobs, and their own – even if, according to their special economic advisers, in the long run everybody will be better off. If they get carried away with liberalising trade, they may find that shutting down a car plant in a marginal constituency puts 3,000 of their key electors out of work. Next election, they too are out of a job.

It's not just jobs, either. Trade has implications for the environment, for culture and for the way societies work, all for which politicians have a keen sensitivity. Unlike economists, politicians see trade as a zero sum game.

They will only open up their own markets in exchange for others opening up theirs. The big trick for a trade negotiator is to get other countries to open up their markets without giving anything at all away. The European Union's refusal to let the Mauritanians market their camel cheese crumbles into insignificance against the US$350 billion of subsidies the rich world pays its farmers directly. When the Doha round of trade talks collapsed in July

2006, the rich world and the poor world alike declared they would rather have no deal than a deal that would offend their own national lobbies.

To cite just one example, in 2005 alone the United States handed out over $3 billion to just 25,000 of its cotton farmers. These lucky people already had assets getting on for a million dollars a piece. The subsidies helped lower the global price of cotton. To compound the irony, America's international aid organisation gave $20 million to the West African cotton-producing countries, who felt the price pressure most keenly.

The World Trade Organization exists to stop this kind of thing and American cotton subsidies were successfully challenged at the WTO by one of the world's other big cotton producers, Brazil. In 2006, the United States scrapped them. But the Brazilian victory will benefit its own cotton farmers, and not those in West Africa.

In Europe and the United States, agriculture employs barely a fraction of the population, but in the poorest countries, agricultural workers make up nearly three-quarters of the working population. But, as the camel milkers bear witness, the rich world is far from relinquishing its iron grip on the world trading regime.

Out of such simple conflicts, bigger ones grow. As globalisation speeds up and we head deeper into Crunch Time, there's little hope that this will change. On the contrary, the game is getting nastier.

Take that, capitalist scum!

> *Is there anything more ridiculous in the news today than the protests against the World Trade Organization in Seattle? ... [the protesters are] a Noah's ark of flat-*

earth advocates, protectionist trade unions and yuppies looking for their 1960s fix.

Thomas Friedman, *New York Times*

It already seems a long time ago, but at the turn of the millennium, it looked as if global capitalism's glaring faults had created an enormous momentum for change. Remember those anti-globalisation riots?

TV images showed tens of thousands of protesters converging upon the World Trade Organization talks in Seattle; the G8 summit in Genoa disappeared beneath clouds of CS gas. The protesters were made up of all sorts of different people pushing all sorts of different causes: trade unionists claiming that globalisation destroys jobs; environmentalists complaining that globalisation destroys trees and whales, and rioting against GM foods or dolphin-killing tuna nets; development lobbies saying poor countries are being dealt a lousy hand by the global trade; and consumer groups shouting about the loss of choice and the corporate takeover of our lives. You name it, they were there.

It was amazing and encouraging. No one could object to activists' compassion for the world's poor while the environment and the workers of the world are being given the raw prawn by the elite that runs global power politics. For many, even for us – white middle-class sons of the rich West and students of capitalism – the instinct was to cheer on the civil action and encourage the nascent 'global justice' movement.

On the other hand, there was one persistent niggling question: What is it that these people actually want? Do they want to stop trade, abolish private property and restrict the very freedoms that allow us to protest at all?

Twenty-first century issues are more complex and interconnected than any we've encountered before, because everybody involved in the interaction is right, and everybody is wrong. The vast majority of the protesters in Seattle weren't there fighting for a return to Soviet-style communism or state control of resources. Instead, the WTO had become the magnet for a myriad of often-contradictory complaints about the ills of globalisation. And there's truth and error in all of the protesters' claims. The fact is that because the world is now a faster, riskier, more contagious place, the 'unintended consequences' of capitalism become apparent violently, critically and very quickly. And it's these unintended consequences that bring the educated middle classes out onto the streets to be baton-charged and tear-gassed.

So what were they protesting against?

Many would simply label it the 'Washington Consensus', after the Washington-based global economic institutions, the WTO, the IMF and the World Bank, which together have decided on a particular type of economic programme for the world, and one that many don't agree with. The Washington Consensus is how the winners see the world economy. It's doing very nicely, thank you. Helping us grow. Putting more cigars in our walk-in humidors and toys in our kids' bedrooms. So let's have more of the same: more openness, more freedom, more trade. Critics call it 'neo-liberalism' – and they're not talking liberal in a pot-smoking, hippy kind of a way. They're talking about laissez-faire liberals from 200 years ago who thought business should be free to do whatever it wanted, because business creates wealth. Neo-liberalism, reckon the protesters, is gospel at the World Bank and the IMF (institutions originally created to help eradicate poverty and encourage growth). It means they support the economic status quo, however unfair, and force poor countries to open themselves up to cut-throat competition, and they make weak governments pursue policies that promote these goals no matter what the environmental or social costs.

Neo-liberalism is what anti-globalism calls the tough side of free-market economics. Radical businesswoman Anita Roddick, herself something of a walking contradiction, defines neo-liberalism like this:

- The Rule of the Market: liberating 'free' enterprise from any bonds imposed by the government no matter how much social damage this causes.
- Cutting Public Expenditure for Social Services: like education and health care and water supply, in the name of reducing government's role.
- Deregulation: scrapping laws that could reduce profits, including measures to protect workers and the environment.
- Privatisation: selling state-owned enterprises, goods and services to private investors. In the name of greater efficiency, privatisation concentrates wealth into fewer hands and makes the public pay more.
- Eliminating the Concept of 'the Public Good': and replacing it with 'individual responsibility'. Pressuring the poorest people to find solutions to their lack of health care, education and social security, and branding them, if they fail, as 'lazy'.

Whether or not you agree completely with her wording and tone, Roddick pretty much captures the main thrust of the economic policies pushed by the IMF, in both the developed and developing worlds. The initial results have, in some cases, been a great boost to growth.

But as we've turned the corner into Crunch Time, the results haven't been so good. It's time for a new paradigm, say the anti-globalisation protesters.

'Hang on a minute', say the Washington types, putting down their cigars and loosening their watch chains. The world is a complicated place, and when you mess with it the results are unpredictable and chaotic. The best way to avoid chaos and to encourage investment and economic activity is to interfere with things as little as possible – let market forces work their magic. Economic stability and trade liberalisation help growth, and growth is good.

What's more, the statistics say its working. According to the World Bank, the proportion of the developing world's population living in extreme economic poverty fell from 28 per cent in 1990 to 21 per cent in 2001.

But despite this, one thing is for sure: the anti-globalisation protesters are right in claiming that neo-liberal policies, combined with the fragility, contagion and risk of the Crunch Time world, have speeded the 'unintended consequences' of capitalism – things such as the 'winner-takes-all' syndrome, which increase instability and environmental and social destruction. How do we know all this? Well, let's look at an example: the global financial system, that edifice of rules, institutions and speculation that provides the engine for real economic growth around the world.

Money movements and economic catastrophes

Only 2 per cent of the US$1.5 trillion worth of international currency that sloshes around the world every day isn't some kind of gamble. Money markets are the on-track betting in the world's economic racecourse. Trading floors even used to look like the bookies' booths at Newmarket.

Most of this money is controlled by the institutions that manage our pension fund money. They are betting that tomorrow the euro will be worth more than the dollar relative to today, or the pound will be worth more than the yen.

This isn't entirely because bankers and financiers like a flutter. The money is genuinely needed to oil the wheels of growth. Global capital markets provide 'liquidity'. This means that when those with cash think there's money to be made by investing in an opportunity overseas, there are people out there willing to provide local funds in exchange for foreign money. Without this liquidity, opportunities would be lost.

At the same time, globalisation deregulation and liberalisation has amplified the impact of money movements on decisions at all levels, from paying the household bills to international debt negotiations. In the interconnected, competitive global economy, with money hopping borders more easily than rain clouds, the people we have elected to actually run our countries are sidelined as international capital markets rampage over their economies.

Deregulated banking systems, money sloshing around the globe, the growth of derivative markets, hedge funds and the odd billionaire investor, have had two fundamental consequences for the world's economies.

On the one hand, economies that have 'opened up' can see rapid benefits in terms of economic growth and international competitiveness. They become more attractive for the global investors and can attract investment capital and factories more easily than others. This puts pressure on all economies to globalise and liberalise, to take advantage of incoming cash.

On the other, 'opening up' an economy means taking away the controls and regulations that stopped large amounts of money from leaving in a hurry – 'capital flight'. What this means is that when a country has a domestic economic 'event' – say, a run on a bank – globalisation means it rapidly comes to the attention of the international investment community, which demands higher interest rates on the cash it's lending because of increased risk. Higher payments make businesses fail, businesses fail and the currency collapses. So, a domestic drama becomes an international crisis.

In 2001, Argentina's 'economic collapse' made world headlines. It was a direct result of the way capitalism is practised in the 21st century, combined with the Crunch Time nature of the world. Argentina is a textbook example of the problems the global capital flows can cause when the markets turn on you.

Before things went belly up, Argentina was an IMF star. Under President Carlos Menem, it put its hand up to every globalised, liberal, free-market policy the IMF thrust upon it. Menem pledged to keep the currency stable against the United States dollar, and hyperinflation calmed down. He privatised everything in sight – ports, railways, utilities, you name it – and in response, the country got rich. In the early to mid-1990s, it had the highest growth rate in South America.

Unfortunately, having allowed foreign cash to walk in freely, Menem found that the same cash could walk out again, without even stopping to say *adiós*.

The first signs of trouble came with a currency crisis in

Mexico in 1995. Mexico devalued its peso. Devaluation scared international investors away not only from Mexico but from the rest of South America as well. They yanked their money from everywhere south of the Rio Grande, Argentina included. The Argentines did their best – they reassured the international money-men by keeping the exchange rate fixed. But the gap kept growing between the interest payments foreign investors wanted on the money they'd poured into Argentina and what the Argentines could pay. The last straw came in 2001 when ordinary citizens thought the banks were going broke and tried to get their money out all at once.

The country defaulted on $155 billion of public debt. The biggest belly-up in history. And it devalued its currency.

Millions lost their savings and their homes, unable to make payments on United States dollar mortgages. A quarter of the country was tossed out of work – unemployment in cities skyrocketed as high as 40 per cent. Poverty became endemic, and income per person was halved in five years, putting Argentina about level with good old Botswana. No Argentine was left untouched by this crisis. All because nervous fund managers squeezed Argentina too hard, then pulled their cash.

This sad story is repeated in many other countries on the periphery of Western wealth. That's to say – in places where most of the world's people live. Huge gains and high growth have been followed by catastrophic crashes, look at East Asia in the late 1990s. When domestic economic shocks hit, when fraud or bad debts weaken banks, what does the international investment community do? It flings gasoline on the flames by betting against countries in crisis and sticking investment cash on the first flight out.

But remember the Washington Consensus? It says financial crises are short-term blips on the upward curve of global growth. It says the benefits of globalisation must override the inherent human desire for stability. Easy to say from an office suite overlooking Capitol Hill, perhaps. Not so easy in a sewerless *barrio*. The rewards for globalisation are an offer that can't be refused. But at the same time, we find ourselves at the whim of the fickle winds of financial fate. High rewards, high risks. It's the nature of our times.

Globalisation 2.0

There's one man who would be desperately disappointed to see the way things have turned out for the international institutions – but probably not desperately surprised. He was the man who told us that we were all semi-pathological, semi-criminals for working more than fifteen hours a week, and he knew a thing or two about globalisation. Here's his description, just after the First World War, of a lost world:

> The inhabitant of London could order by telephone, sipping his morning tea in bed, the various products of the whole earth, in such quantity as he might see fit, and reasonably expect their early delivery upon his doorstep; he could at the same moment and by the same means adventure his wealth in the natural resources and new enterprises of any quarter of the world, and share, without exertion or even trouble, in their prospective fruits and advantages ... But, most important of all, he regarded this state of affairs as normal, certain, and permanent, except in the direction of further improvement, and any deviation from it as

> aberrant, scandalous, and avoidable. The projects and politics of militarism and imperialism, of racial and cultural rivalries, of monopolies, restrictions, and exclusion, which were to play the serpent to this paradise, were little more than the amusements of his daily newspaper, and appeared to exercise almost no influence at all on the ordinary course of social and economic life, the internationalisation of which was nearly complete in practice.
>
> John Maynard Keynes, 1919

Replace the phone with a servant and the story could move back a couple of thousand years to Rome. Trade has been a feature of human life that pre-dates history. Obsidian was being exchanged long before people began to farm. Trade can furnish grave goods but it can't guarantee the future.

In the twenty years after Keynes wrote this, the world fell swiftly into another massively destructive war. Those conflicts underline the fragility of what we feebly imagine to be the irreversible progress of globalisation.

Towards the close of the Second World War, Keynes was on Britain's side of the negotiations to determine how the post-war world would be set up – the design of the post-war institutions – the World Bank, the IMF and the General Agreement on Tariffs and Trade, which morphed into the WTO.

On the other side of the table was America, which held all the cards and all the money. Unsurprisingly, the hand that had the most influence in shaping the institutions was America. And, as we shall see in the next chapter, this influence has entrenched the system that has given rise to globalisation 2.0.

Still, the way the Washington institutions were set and the way that the post-war world has developed tells two different stories. On the one hand, America undoubtedly wanted the world built to its own design, and it's done a lot of questionable things in the half-century in which it has ruled the roost. Like the British Empire, it stands accused of arrogance and self-interest. But at the same time, look at what's happened in the world since the end of the Second World War. In 1945 America was even more pre-eminent than it is now, as Japan and Europe had been laid to waste. Rather than grind them into the dust, America helped those countries create democracies and build up their economies – no doubt to furnish markets for its own products – but nonetheless, it was an enlightened self-interest that went beyond the treatment meted out to Soviet Russia's conquests on the opposite side of the Iron Curtain.

Into the 21st century, too, America's own growth and stability relies on the growth and stability of the whole world. But that doesn't mean that powerful players won't always want to tilt the playing field in their direction.

Globalisation is a two-way street – although the contrast between Den Fujita's success in Japan and Nancy Abeiderrahmane's failure in Mauritania doesn't make it look that way ... at first glance. But the stories don't end there.

Den Fujita retired as chairman and CEO of McDonald's Japan in March 2003, as the company slipped into the red to the tune of $20 million. Japan's economic woes, combined with a series of protests against the fast-food chain's meat handling and animal-rights record, forced the company to cut prices to such an extent that it was pretty much making a loss on every burger it sold –

and because they were so cheap (and, at the end of a ten-year recession, people were feeling so poor) they sold a lot of them.

The company tried all sorts of tactics to dig itself out of the mud, broadening the menu, changing its advertising, but, say commentators, McDonald's Japan is just a victim of a more powerful force than globalisation: fashion. As one teenager Japanese noted: 'It used to be McDonald's every evening after school, but not now ... We used to drink coffee, or maybe eat something, but the main thing was knowing that everyone would be there. McDonald's just isn't somewhere we want to meet anymore.'

As for Abeiderrahmane ... at the time of writing it looked to be several more years before she would be successful in accessing the wealthy Western markets with her camel cheese. 'Please note: to date these products are not available for export', says the company's website, but her attitude was best summed up by the sign on her office wall: 'Don't EVER give up'. Good advice, perhaps for all of us, as we head deeper into Crunch Time.

CHAPTER 4

Envirocide

The collateral damage of being human

How long can man go on overpopulating this planet, destroying its topsoils, slashing off its forests, exhausting its supplies of fresh water, tearing away at its mineral resources, consuming its oxygen with a wild proliferation of machines, making sewers of its rivers and sea, producing industrial poisons of the most dreadful sort and distributing them liberally into its atmosphere, its streams and its ocean beds, disregarding and destroying the ecology of its plant and insect life? Not much longer I suspect. I may not witness the beginning of the disaster on a serious scale. But ... much of the damage that has already been done is irreparable in terms of the insight and effort of any single generation.

George F. Kennan, 1969

Holidays. Adrian was back from France, and moaning. 'It was horrendous', he said. 'We drove to a barn we'd rented in the middle of nowhere. Wife, two kids and hours of driving in blast furnace heat with no air con. It was 41°C on the hottest day, I kid you not.'

Europe was in the middle of another mercury-busting summer, and Adrian was regretting driving his family in their unair-conditioned old banger across the continent's melting roads.

'As we drove back in the heat and traffic, I said to Linda, we're getting a new car, with air-conditioning. One of those big four-wheel drives – we do live in the country.'

'So you're going on at me about how damned hot it is and how the heat spoiled your holiday, but you want a huge bloody SUV that spews carbon dioxide into the atmosphere and guzzles a tank full of gas just pulling out of the drive. Hello. Join the dots, please.'

Mike pointed out that thanks to global warming, British researchers reckon that by 2040 every other summer in Europe will be a blast furnace.

Not one to shirk a challenge, Adrian laid out his position.

Firstly, the world may be getting warmer, but in South-East England it makes a very pleasant change from the light drizzle that usually passes for summer.

Secondly, if the heat wave is a symptom of climate change, it's a long bow to draw between the travelling comfort of the Monck family and global warming. Many things may be contributing to climate change. Carbon dioxide emissions, OK, but also natural variations in climate, methane gas from cow farts and landfill, air travel and all sorts of imponderables.

Thirdly, even if buying a bigger car with air conditioning would contribute to global warming, it would be a tiny contribution compared to the personal benefits of riding around in a refrigerated Range Rover.

Mike had to admit that Adrian was right. But at the same time, they both knew he was wrong.

Heating up to cool down

The hot French summer produced another example of a kind of Crunch Time catch-22. In 2003, thousands of mainly elderly people died from heat stress in one of the hottest summers the country had experienced. The next year the French had their electric fans, refrigeration units and air conditioners at the ready. So the heat increased demands for energy. About a third of all the water used in Europe goes to cool down electrical generators, including those powered by coal, gas and nuclear fuels. Fossil fuels, of course, shoulder some of the blame for global warming.

The French get nearly all their electricity from nuclear power, increasingly cited as the answer to global warming, and hot summers in turn. There are nearly 60 nuclear power stations in France, three-quarters of them by rivers where they can use the water for refrigeration and then dump the heated wastewater back into the supply. Too much hot water and the river dies. In July 2006, it got so hot that the rivers ran too low and too warm to be much good for cooling. Reactors had to run at lower capacity and the French were forced to buy in electricity from neighbours to keep the power supply from going off. One person turning on the air conditioner doesn't make much difference, what matters is when everyone does it.

This is one example of an economic problem that lies at the root of many Crunch Time environmental issues.

The tragedy of the 4x4

Back in the 1830s, Oxford professor William Forster Lloyd asked himself why the cattle on a common were so: 'puny and stunted? Why is the common itself so bare-worn, and cropped so differently from the adjoining inclosures?'

Lloyd's observation was picked up in the late 1960s by a whacky ecologist named Garrett Hardin (of whom more later) who labelled it 'The Tragedy of the Commons'. The Tragedy shows how we can all act rationally to destroy a commonly held resource. Here's how it works. There's a common, a pasture, where anyone can graze cattle for free. Because it's free, each cowherd will try and keep as many cattle as possible there. This may work well enough for centuries because war, rustling and disease keep the numbers of man and beast below the carrying capacity of the land. Finally, however, stability arrives, and with it, the tragic logic of the commons.

So, each cowherd wants as large a herd as possible out grazing on the common. What are the pros and cons to the individual shepherd of sticking an extra cow out to graze? They each do a little mathematics to figure it out.

- **The pros** – As the cowherd pockets all the cash from selling an extra animal, the positive utility is almost plus one cow for each cow sent to graze.
- **The cons** – After a point, one extra cow nibbling away at the common will cause overgrazing. But since the effect of overgrazing hits all the cowherds, the negative utility for an individual cowherd is just a fraction of minus one.

Add the two partial utilities together, and it's a no-brainer. Stick another cow on the common! And another! Every cowherd using the common comes to the same conclusion. And therein lies the tragedy. The pasture is overgrazed and destroyed. Everyone ends up screwing things for everyone else.

The 4x4 plague is another tragedy of the commons. SUVs, as Adrian so aptly put it, are a great boon to personal transport. They are big, safe and comfortable, and they make their owners feel superior. But from the wider community's point of view, they're a disaster.

Firstly, they might be safer for those driving inside them, but for everyone else on the road they're a major hazard. SUV drivers can't see out the back, so stories of toddlers run over by their own mums on their own driveways are a staple of tabloid newspapers. And they are difficult to manoeuvre, particularly for the demographic that seems to be attracted to this particular type of vehicle (middle-class, inner-city family drivers, although Adrian hastens to point out that the Moncks live in the country), so their accident rate is higher than the average saloon.

More than that, because the incidence of SUV ownership is on the increase, others on the road feel unsafe if they are driving smaller cars, so pressure to buy into the SUV myth is intense.

And, as Mike pointed out, SUVs are an environmental disaster. Half of the oil the world uses goes on transportation. SUVs are far heavier than other passenger vehicles, and take a correspondingly large amount of resources to create and destroy. They chew up fuel faster and emit more pollutants than other vehicles. And because they are so big, and safety concerns are so strong, they boost consumer resistance to the development of smaller, lighter, potentially much more environmentally friendly vehicles.

SUVs are a tragedy of the commons because their owners, who get all the benefits, don't incur all the costs. The costs of wider SUV ownership are imposed on wider society – the commons. But we, the wider community, have no say in this because (although some councils are talking about charging higher road tax for heavier passenger vehicles) it's not, at the moment, recognised in the price to the SUV driver on the road.

The SUV is just one example of the commons tragedies that are plaguing our world. In Crunch Time, we'll all have to pay the environmental costs of the behaviour of our neighbours – and the bill is large. Humanity has survived climate change before, but there's a difference between surviving and prospering. Before we get to that, however, we ought put things in perspective.

Nothing new there then

One millennium ago, when the first Viking settlers' steered their way into the fjords of Iceland, they were entering a virgin land. After seven brutal days on the North Atlantic, crammed into single-masted, square-rigged open ships barely 60 feet long, the immigrants stumbled out onto an island that was a brand new natural idyll, still packed in God's box. They were greeted by uninhabited, pristine wilderness. Lush forests of birch, willow and alder stretched from the base of the island's volcanic mountains to its dark shoreline.

The new country offered land and freedom. With neither kings nor lords to rule them, the Viking settlers made their own laws.

Six hundred miles from Norway, their homeland, the new Icelanders came prepared for self-sufficiency. They brought their households, their loved ones and their livestock. When they arrived, the only mammals living on the island were seals, arctic foxes and field mice. The Vikings carried cattle, sheep, pigs, goats, horses, cats and dogs with them.

But the thing they needed most was wood and timber.

Timber built ships, homesteads and fences. Wood was burned in hearths, and provided the charcoal for smelting bog-iron.

The first people to arrive on Iceland felled the biggest trees to provide frames for their farmhouses and repair their damaged ships. Successive settlers burned forests to create pasture when the grasslands were used up. As Iceland's population grew and became more industrious and prosperous, overeager herdsmen quickly overstocked and overgrazed the treeless valleys, and hacked away at the remaining trees. With no system for conserving the island's forests, the amount and quality of timber the island could provide was quickly exhausted.

With no one managing it, the island's supply of wood was finite, and the Icelanders' appetite for it practically infinite. Like today's SUV owners, individual farmers weren't thinking about the greater good or the future of their people as they chopped and cleared.

Barely a generation after settlement, Iceland's trees were gone. Good timber had to be imported, making longships too expensive to keep seaworthy. By the 1100s barely any Icelander owned an ocean-going boat, and that meant no travel. No travel meant no trade.

It also meant no deepwater fishing. Iceland, marooned in a vast, fertile ocean stocked with shoals of fish, schools of whales and alive with sea mammals, had no fishermen, just land-locked, subsistence farmers. They were subsistence farmers because without wood it was impossible to fence off grasslands to make hay to keep animals alive through the long hard winters.

By destroying their forests, Iceland's Vikings lost the independence they'd cherished for generations. They became dependent on Norway for timber, trade and ships, and Iceland became a remote outpost, ruled for centuries by Norwegian kings. Now it's just a history lesson to illustrate the value of environmental resources to our great game of *Civilization* – economically, socially, politically, however you want to define it.

Icelanders aren't the only ones who messed things up by destroying the ecosystem around them. The same thing happened on Easter Island and it's happened too in other closed environmental systems into which man has inserted himself.

Surely, no modern land could repeat these dumb, destructive mistakes? Surely not somewhere as innovative, vibrant and resourceful as, say, Australia?

For 'timber' read 'salt'

Looking back through the fog of a thousand years to the Icelandic Vikings, it's still possible to imagine their sense of optimism as they landed in a new country after seven days sailing across the northern seas. Not so the wretched Aussie colonials of the First Fleet who endured not seven days but seven months crossing forbidding oceans only to pitch up on a harsh and hostile continent. For them there was only the prospect of years of labour far from everything they had ever known.

Still, human energy and ingenuity being what it is, many of the criminals who had been transported down under turned their lives around and became respectable and well-off farmers, merchants and politicians, effectively duplicating the European lifestyles and social structures they had left behind. They weren't to know, of course, that by doing this they would lay the groundwork for the 21st century's equivalent of the Icelandic tragedy of the commons.

In 1789, while France was guillotining its way through a revolution, Australia's first farmer, James Ruse, was serving time for breaking and entering. Granted a dozen acres in Parramatta on release, he began stripping his land of native flora and fauna and planting grain and vegetables. He was paving the way for generations of pastoralists to do the same. But Ruse, and his enthusiastic successors, began the process of fundamentally altering the ecological equilibrium of their new home.

Australia is the world's flattest, driest continent. Its native plants are thirsty. When it rains, they soak up pretty much all the water that falls from the sky. When a farmer clears the land and replaces bush and scrub with scrawny European plants, rainwater leaches through to the ground below, dissolving the vast amounts of salt that have lain as crystals in the soil since time immemorial. And there's not just a sprinkling of the stuff. Under each square metre of Western Australia's wheat belt lurks up to 120 kilograms of salt.

As excess rainwater drains through the soil, it takes the salt with it. The more salty water goes down, the higher the water table rises – only this time it's the saltwater table. Saltwater destroys crops, rots buildings, rusts rails, ruins roads and costs millions.

The technical term is 'dryland salinity' and it's one of the greatest challenges currently facing Australia's rural community, with more than 17 million hectares (an area double the size of Tasmania) threatened by salt. It's another utterly predictably, utterly unforeseen tragedy of the commons – James Ruse enjoyed the benefits of his cleared farmland without experiencing the costs of the dryland salinity that it would create.

Yet this is just one tragedy in thousands caused by

exactly the same cause: people. How on earth did we get here?

How farming filled the world

About 12,000 years have passed since humans first began cultivating crops and raising animals. There are many mysteries remaining as to how exactly it came to pass and why anyone actually bothered. Why was it that suddenly, after aeons of happy hunting and gathering, people decided to settle down and become farmers almost simultaneously in different places across the world?

Was it environmental change? A case of goodbye glaciers, hello horticulture? Unlikely. Was it that people suddenly thought it was sophisticated and clever to grow stuff and look after sheep and cows, rather than pick wild berries and hunt? Probably not.

No, theorists reckon it was something altogether different. As we discovered to our own dismay, when a hapless young man suddenly finds himself in the family way, he is forced by social and economic pressures to settle down. And that's exactly what happened to *Homo sapiens* as a species.

As we became more successful and our numbers began to grow, the regular, carefree existence we had lived as cave-dwelling hunters and gatherers just wasn't enough. Our ancestors, far from living in balance with nature, had successfully engaged in a millennial slaughter of big game that eventually forced us into farming. The newer lifestyle of cultivation and husbanding animals allowed us to provide for our ever-increasing numbers.

But we still don't know if it was such a good thing, because the benefits of becoming a son of the soil aren't so clear-cut. All the available evidence says that hunting and gathering makes for a pretty laid-back existence. The Hadza people in Tanzania are probably the last foragers in Africa. Their way of life is the closest thing we have to seeing what life was like before agriculture. According to one anthropologist who lived with the Hadza, they 'meet their nutritional needs easily without much effort, much forethought, much equipment, or much organization'. Life was spent hunting, foraging and gambling. Some anthropologists call hunter-gatherers the 'original affluent society'. Progress – in this case, from hunting and gathering to farming and settling – isn't always forwards.

Compared to farmers, hunter-gatherers have a wider range of foods at their disposal, and so face arguably lower risks of famine. Their diet is more diverse and healthier, and generally they consider their way of life to be rather better than that of crop growers or pastoralists. If you thought farming was a natural progression on the long, slow, upward path of man, you'd be wrong. Archaeologists distinguish the remains of the first farmers because they are shorter and weedier than their hunter-gatherer predecessors. Nor is there much evidence that agriculture is more efficient in terms of basic yield per hour of toil than foraging.

It's also less healthy. Close proximity to domesticated animals allows diseases to jump species. Pigs and ducks gave us flu, horses the common cold. Measles is a form of canine distemper, courtesy of man's best friend. Converting forests into farmland created ideal conditions for malarial mosquitoes. Irrigation brought waterborne diseases like bilharzia.

Farming sustained more people, but it did so at the cost of poorer health. Dining on staples like wheat or rice or maize day after day made for a dull diet low on vitamins and protein. Even so, those crops enabled our ancestors to feed ever more ancestors. Hunter-gatherers on the move could only raise children every three or four years. Farmers could grow their own workforce, year after year. Those growing populations needed more food, and the only way to provide it was agriculture. More farming needed more farmland, more farmland meant clearing forests, digging irrigation channels and so the cycle began again.

Anthropologists call it the Neolithic Revolution. From our 21st-century perch, we can see it for what it really was: the beginning of the rat race. And eventually it forced itself on the people it helped create – everyone. Thanks to farming our days were duller, more arduous, and our rewards less varied than when we had hunted. But there were too many of us to go back.

The next time you read in the news that we're working harder and harder for less and less, take solace: it's hardly unprecedented.

Agriculture produced the surpluses sufficient to provide for the priests, scientists, artists, politicians, scholars, and so on whose activities are what we collectively think of as 'civilisation'. In the millennia since, the world's population has grown to over 6.5 billion people. We all might be working harder than we would as hunters and gatherers, but our ability to reproduce has increased rapidly – particularly in the last 300 years. Meanwhile, we have built villages, towns, cities and all the good things that make up that civilisation.

Why we hate terrorists, not global warming

Farming, and the change in the world's geography that it's caused, is only 12,000 years old. Twelve thousand years isn't very long – it's only about 500 generations. Long enough to grow taller and invent silicon implants, but not long enough to invent an effective cure for baldness or temper our most ingrained patterns of behaviour, those we learned as hunter-gatherers. These behaviours include putting emotion before reason, confidence before realism, classifying others into 'us' and 'them', competition, contest and display. No longer wandering the earth clad in animal skins, we nonetheless carry this psychological baggage with us when we're driving, or in meetings, or online, or watching football. Our technology might be wireless but our brains are wired: to live in groups, forage, to kill furry things and occasionally each other.

As human beings we operate on mental principles that discount future uncertainty, by putting today over tomorrow at almost every opportunity. Celebrated economist John Maynard Keynes put it this way: in the long run, we're all dead. Naturally, this individual truism is a recipe for collective disaster.

The problems we're facing in Crunch Time are a direct result of the mismatch between our success and power over the world around us, and the world's ability to keep providing us with hospitality. In 2004, four of the world's leading scientists came up with a name for it – the 'Anthropocene' – the geologic epoch in which humans are a significant and sometimes dominating environmental force.

In other words, we're guests in danger of overstaying our welcome.

Technologists vs. tree-huggers

Perhaps the most important aspect of the greenhouse-ozone-acid-rain complex, and of their as-yet-unknown cousin scares which will surely be brought before the public in the future, is that we now have large and ever-increasing capabilities to reverse such trends if they are proven to be dangerous, and at costs which are manageable.

Julian Simon, 1994

One fact clear to all concerned is that the natural world around us is changing radically and rapidly, and that this change will only get faster. Those changes will be the cause of much grievance and conflict, and they are intimately connected with the other issues we discuss in this book: money and the markets, population and poverty, economic migration, and the rest.

Put simply, there are two approaches to the environmental issues that will emerge during our lifetimes, each put forcefully and often set against the other. Shrill rhetoric and a gulf of misunderstanding between the proponents of each makes them seem mutually exclusive and in desperate conflict with each other, but in actual fact they are complementary and interchangeable, and our only hope is the intelligent application of both.

The first approach is that of the Technologists, with special appearances from the Economists and the Futurists. The Technologists feel that we've made an unbreakable pact with science and invention – and since they got us into this mess they will have to get us out of it. This side of the debate thinks that the only way to accommodate the burgeoning numbers of people on the planet and their material desires is to apply technology to solve the problems that development creates.

A couple of hundred years ago, the economist and clergyman Thomas Malthus – of whom more later – was predicting that human population growth would outstrip our capacity to produce food by the middle of the 19th century. The global population then was about a billion people, less than a sixth of what it is now. Malthus, say the Technologists, was wrong. We can now feed the world because we have pesticides, nitrogen-fixing fertilisers, and combine harvesters the size of apartment blocks. We now husband over 3 billion cattle, sheep and goats, and apply scientific fishing techniques to trawl the oceans. There are many hungry people in the world, but the problem is often not the ability to produce food, but the ability to get it into the mouths that need it. Our growth in numbers is a remarkable achievement for humankind and a point on the scoreboard for the Technologists.

In the same vein, the Technologists point to a string of false prophecies that has come from the other camp. In the early 1970s, the Club of Rome, a ponderous group of European elder statesmen produced a doom-laden tome called *Limits to Growth*. They reckoned that humanity would soon run out of resources. Here are some of their predictions:

1981 – Bling never even gets off the ground because gold supplies have run out.

1985 – The temperature might be going up but the mercury isn't moving. There's none left.

1990 – French café owners mourn the depletion of zinc.

1992 – There's no more gas to guzzle – petrol supplies are exhausted.

1993 – Copper, lead and natural gas – all gone!

It seems they were a little off the mark. Since the 1970s, known reserves of all these commodities have increased, substitutes have come along for some and prices have fallen. The Technologists reckon this is because human society is ingenious, and when resources become scarce, people find new ways to extract them, or switch to substitutes.

The Technologist position is buttressed by the fact that the rich world has been getting relatively cleaner and has suffered less environmental destruction over the last 30 years, while the poor world has only got dirtier. Why? Because, they say, more economic growth means more technology, more money and more time to deal with environmental issues, and so a cleaner environment. Best encourage the poor world to catch up with us. Free trade and market forces, go, go, go.

Technologists view the future as a global game of whack-a-mole. Environmental crises pop up, only to be bashed on the head by the swift hammer of technology. Take the dryland salinity problem we mentioned earlier. In 2001, Professor Eduardo Blumwald from the University of California came up with a tomato plant that could grow in 40 per cent salt water. Problem solved. Unfortunately no local growers wanted to grow it, so the original plants were thrown away. Technology, from biotech to nuclear power, has 'image' problems.

Hang on, say the Tree-huggers. The problem isn't too little development, but too much. Surely, an ounce of prevention is worth a pound of cure? Our faith in science and technology as a cure for future woes is actually the cause of all our troubles.

Yes, we can feed more people now than we could, but at what cost? We spray our farmland with 3 million tons of pesticides a year. We produce more nitrogen than the whole global total derived from natural processes. These things are poison. About two-thirds of the oceans' fisheries are being trawled to destruction. The portion of the earth being farmed may have grown to an area about the size of South America, but most of the pasture is grazed at or above capacity. Our yields depend on oil-based fertilisers. We have lost about a fifth of the world's topsoil, a fifth of its agricultural land, and a third of its forests over the past half-century. We have changed the composition of the atmosphere profoundly. We've added to the carbon dioxide that's helping to heat the earth up, decimating habitats and native species around the world, and causing goodness knows what kind of problems further down the line.

Technology can't save us from ourselves, say the Tree-huggers. We need to save ourselves from ourselves, more than that, we need to save the natural world. Their vision of the world is one in which we awake from our technologically driven dream world, strap on our sandals and get back in touch with our true selves, living in harmony with nature, and ultimately scaling down. Yes, fewer baths and all that kind of stuff.

Tree-huggers reckon that when we're living at the end of nature it's unsurprising that our violent lifestyle proves violent in return. For many hard-core huggers, the issue isn't that the earth will bite back at human society – human society probably deserves it anyway. The problem is the simple irreversibility of the loss of the environment.

Once rainforests are gone, they're gone. Once a tract of land is concreted over with a car park, it's destroyed, once and for all. Once a species is extinct, like the Norwegian Blue in Monty Python's 'dead parrot' sketch, it has expired and gone to meet its maker – kaput, finis.

Tree-huggers take as their starting point that life on earth, human and non-human, has an intrinsic value in and of itself. That the existence or not of a particular species or wilderness is important because it's important, not because it necessarily provides man with something nice to look at, eat or to cure his ills with. What this means is that the environment comes first, and humans have to play by nature's rules. These people reckon that the Icelandic Vikings deserved a good hiding from Mother Nature for putting their needs first. And they'll be there laughing when the last crop-duster crashes into the last computer. Cheerful stuff.

So they're poles apart, the Technologists and the Tree-huggers. But there's a movement that's trying to bring the two poles of opinion closer together. Step forward unlikely superheroes, the Economists. One fundamental problem that we can change, say these optimists, is that decisions about production and consumption don't take into account environmental costs because the price doesn't reflect it. Remember the tragedy of the commons: Adrian's SUV.

Our society is organised around economic principles that for the most part fail to take into account the environmental consequences of our choices. No matter what our intentions, if we want to achieve anything in the material world at least, it's extremely difficult to live an ecologically balanced lifestyle. Our everyday economic framework and our habits and behaviours mean that we live as if we were nothing more than consumers from birth until death, almost completely ignoring our impact on the wider world around us.

Environmentally aware Economists are busy trying to come up with new taxes or incentives that will put a value on the environment, and give ownership of, and responsibility for, 'the commons' to all of us consumers. Tree-huggers despair at the very idea that we should put a price on purely environmental goods, such as clean air and water, abundant fish stocks, or a species saved from extinction. And meanwhile, politicians are busy trying to make sure that their constituents (that's us, too) don't have to pick up the tab.

The 4x4 factor

Economists argue that we take environmental decisions by default every day – so unless we try to value all the components of our decisions, we'll simply be making poor, uninformed choices. This may be so, but even with the benefit of cutting-edge environmental valuation techniques, we're still making poorly informed choices, only now they have a veneer of scientific acceptability, and conform with the prevailing economic myth that drives so many of our decisions in this day and age. Sometimes this makes them even more dangerous.

Take the high-profile issue of global warming, for instance. It's fiercely complicated both scientifically and economically, but the super-short version is as follows: human activities (farming, forest clearing, transportation and industrial activity in particular) cause the emission of carbon dioxide, methane and other greenhouse gases – those that make the atmosphere trap heat. There's no doubt that the amount of these gases in the atmosphere has increased over the last hundred years, and that we are the cause. This much is uncontroversial. The earth's climate has been warming by about 0.7 °C over the past century. Forget French holidays, say climatologists, it's the global averages that matter.

According to the Geneva-based World Meteorological Organization, 2006 was the sixth-warmest year on record, and only marginally cooler than 2005 and 1998, which are the two hottest years on record. Six of the ten hottest years have been in the 2000s, and all ten have been since 1994.

And there's other evidence, aside from temperature figures, to suggest the earth is heating up. Less snow on mountaintops, melting glaciers, changes in rainfall patterns and an increase in the intensity of hurricanes, possibly due to warmer seas, are also confirmation of global warming.

Sea levels are creeping up by around 1.5 millimetres a year. They have risen by at least 20 centimetres since the late 19th century.

Because of the many uncertainties in climate change science, a UN body was established in 1988 to collate and check the data and its findings are a major influence on politicians deciding what to do on climate change. That body is the Intergovernmental Panel on Climate Change (IPCC), which in a 2001 paper said temperatures were set to rise by 1.4–5.8°C from 1990 to 2100 – with potentially devastating consequences.

For the IPCC scientists there's little doubt that warming is happening, it's caused by human activities, and its consequences will be serious. Mike's already seen it. In September 2003, when water restrictions were introduced for the first time in New South Wales, then premier Bob Carr declared that it would be remembered as the first time global warming touched Australians' lives (Bob hadn't been with the Moncks on holidays).

How do scientists know all this? Through historic and current observation and through computer modelling that tries to take into account the influences on the earth's atmosphere that combine to make up our aggregate climate.

'The science on climate change is overwhelming', the editor of *Science* magazine noted in 2001, a 'consensus as strong as the one that has developed around this topic is rare in the history of science'. Accusations of alarmism have meant the message has taken decades to get through. Even when it has, scientists have faced scepticism and indifference.

The scepticism comes when you ask what we should do about it, and who this will affect.

Stopping global warming

> *An increase of two or three degrees wouldn't be so bad for a northern country like Russia. We could spend less on fur coats, and the grain harvest would go up.*
>
> Vladimir Putin, Russian President, 2003

Winter 2006 in Moscow was the hardest in a quarter of a century. Russians might well wonder if climate change is dealing them a winning hand. When it comes to reducing greenhouse gas emissions globally, agreement has been the first problem.

But it's not the only one. Monitoring emissions on a country-by-country basis isn't easy. Britain stands accused of a massive 92 per cent under-declaration on its emissions because of rubbish. Methane seeping out of landfill just isn't being owned up to. So even if you get agreement, getting people to stick to it is another matter.

There's more. Europe, the United States and Australia are still some of the biggest offenders when it comes to emissions, mainly through burning fossil fuels. By cutting back they could dramatically lower the amount of carbon dioxide being chuffed into the atmosphere. But do we want to? We want air con to relieve the summer heat that kills the elderly. We baulk at big taxes on already expensive fuel that funds our trips to superstores and keeps their shelves stocked. We worry about the dubious benefits of nuclear power. We fret about airline travel but need it to keep families together across continents, as Mike and his wife know only too well.

The Economists can't really challenge the science. But they can argue about what to do about it. Before we act, they say, we have to assess the costs. And who better to do that than the people who end up paying out – the insurance industry. In 2002, a UN group chaired by the world's biggest health and life insurer, Swiss Re, concluded that worldwide economic losses due to natural disasters were doubling every decade and, within the next ten years, could reach US$150 billion annually. In insurance, the customer pays.

Why we spend more on terrorism than global warming

Our reticence to actually do anything about global warming comes from sound psychological factors borne of our hunter-gatherer ancestry. Firstly, we're mostly concerned by what other humans do, not by atmospheric gases. Understanding what others are up to – what they know and want, what they are doing and planning – has been so crucial to the survival of our species that our brains have developed an obsession with all things human. When psychologists observe us they see us thinking about people's intentions: remembering them, discussing them, predicting them.

This obsession with intentionality means we worry more about terrorism than car crashes. The attacks on the London Underground on 7 July 2005 killed the same number of people who die each week on Britain's roads. Road deaths are accidental, but terrorism is intentional, and the smallest human act of violence captures our attention in a way that the biggest accident does not. When Hurricane Katrina swept through New Orleans, it wasn't the force of the storm that had us all talking, it was the ineptitude of the Bush Administration. George W. Bush didn't make the hurricane, but we had to blame someone – the President – not something, i.e. the weather ...

Our adaptive strategies make us peculiarly ill suited to mobilising to tackle environmental issues. The war on terror can be personalised with names and faces – 'bad guys' – the war on global warming can't.

At the same time, global warming has no moral content – it doesn't force us to entertain thoughts we find repugnant or sacrilegious. It fails to offend our moral sensibilities. When we feel insulted or disgusted, we generally do something about it – moral emotions are the brain's call to action. Global warming doesn't sicken, anger or shame us, and so we don't feel compelled to react against it as strongly as such a major threat to our society's future as say fox hunting or gay marriage.

All human societies have moral rules about food and sex. None has a moral rule about the composition of atmospheric gases. The fact is that if climate change were caused not by power plants but by paedophiles, millions of protesters would be massing in the streets.

Thirdly, global warming doesn't trigger our concern because we see it as a threat to our sunset years, not to sunsets period. Our brains are engineered to scan the environment for clear and present dangers. We are responsive mammals and we can duck a football in a park in just a few thousandths of a second, but we can duck issues even faster.

But just a few million years ago, the mammalian brain learned a new trick: to predict the timing and location of dangers before they actually happened. Our ability to preempt that which is beyond the horizon is one of the brain's most stunning innovations. We wouldn't have tooth brushing or insurance without it.

We haven't quite got the knack of treating the future like the present because we've only been practising for a few million years. If global warming had human followers who plotted to destroy our society's infrastructure, the security services would be tracking its every move.

The human brain is exquisitely sensitive to changes in pressure, weight, size, sound, light and temperature. But if the rate of change is slow enough, it goes undetected. Because we barely notice gradual alteration, we accept big changes that we would reject if they happened abruptly. The density of traffic on our roads has increased dramatically in the last few decades, and citizens have tolerated it with only the obligatory grumbling. Had that change happened on a single day last summer, it would have brought Britain to a grinding halt.

Tree-huggers despair that global warming is happening so fast. In fact, it isn't happening fast enough. If we could all go on holiday to 2050 then we would assuredly be willing to make some of the changes necessary to stop the future.

We're the descendants of short-lived people who hunted and gathered, whose greatest threats were large predators and one another. We respond to terrorism with force, road accidents with fatalism. Global warming fails to alarm us, and like the frog in the pot with the heat turned up we may not notice we're boiling alive until it's too late.

The sustainability mantra

The Economists and the Tree-huggers put themselves at two poles of the spectrum, but there is a word that bridges the gap. And you've probably heard it.

Pick up any environmental report from any international body (the UN, the World Bank, development agencies, you name it) in the last ten years, and the one word that will repeatedly rear its head is 'sustainability' – as in the 2002 Johannesburg Summit on Sustainable Development. Sustainability is a buzzword, for sure, with as many different meanings as it has mentions, but at its best, it's where the Technologists and the Tree-huggers meet and shake hands (with a fair degree of mutual suspicion, to be sure).

Sustainability means living a considered life – thinking about what we do rather than just doing it, and taking into account the discount rate that we're levying on the future when we make our decisions.

Sustainability also means being answerable to the future. It means not taking so many fish from the sea that there will be none left tomorrow. It means that while we're searching for energy supplies that won't heat the global climate, we take it easy on car journeys, or subsidise electricity that comes from windmills rather than coal-fired power stations.

It means constructing government policy that redresses some of the destruction inflicted upon our natural surroundings by our violent economic system, by our rapid population growth, and by our general thoughtlessness. It means measuring the environmental damage we make in our everyday lives and figuring out ways to stop it.

It means that if we're confronted by an environmental crisis we look beyond the immediate issue towards a way of ensuring that the solution we implement doesn't cause greater problems tomorrow than it did today.

It means changing our value system so that consumption of non-renewable resources is priced properly. It means looking beyond today, using and polluting less, and leaving some of what we've got for our children. If the Icelandic Vikings had understood the concept of sustainability, how the course of their history may have been altered. If we don't come to grips with it, the consequences for humanity as a whole will be equally significant. It is, to use an already well-worn phrase, Crunch Time.

Protecting the commons

We started this chapter with the tragedy of the commons and its message of unconscious destruction. But while Vikings were chopping down Iceland's last tree, high up in Vispertal in the Swiss alps the canny peasants of Törbel were figuring out how to preserve their shared mountain space together. Written legal documents dating back to 1224 provide information on the types of land tenure and the rules used by the villagers to regulate the five types of communally owned property: the alpine grazing meadows, forest, wasteland, the water supply, and the paths and tracks.

Written regulations specified in 1517 that no citizen could send more cows to the alp than he could feed during the winter. The regulation is still enforced.

If William Forster Lloyd had got out of Oxford more he might have made it to Beverley in Yorkshire where 'Pasture Masters' looked after the commons. The current law that governs their membership goes back to the 1830s, exactly when Lloyd was writing, and their job then, as it is today, is to look after a shared resource for the wider community's benefit.

The message is that in order to be answerable to the future we don't necessarily have to go barefoot everywhere, chain ourselves to trees and hurl ourselves out of dinghies against oil rigs in the North Sea. But we should at least be able to identify where our lifestyles are inconsistent with our values and do something about it. There are forces that act on our way of life – economics, government regulation, social norms – that can ally with personal behaviours to mobilise against things like climate change.

The Moncks never did buy their 4x4. Adrian says it was the insurance.

CHAPTER 5

Powering our Lifestyle

Doom, gloom and optimism in the energy markets

Adrian's SUV conundrum is a common theme among environmentalists, policymakers and TV writers. In an episode of *The West Wing* entitled 'The Hubbert Peak', Josh Lyman, Deputy Chief of Staff, goes out to buy a new car. Representing the most liberal White House since vestibule sex was the done thing, he is duty bound to opt for the greenest of options – a Toyota Prius hybrid car, which has low carbon emissions.

Problem is that the waiting list for a Prius is a mile long ... and Josh's eyes get to wandering. They wander to a nearby Humvee, an enormous SUV that looks like an awful lot of fun, and Josh decides to take it for a test drive. While talking on his mobile phone, he crashes this enormous car into someone else's new Prius, causing all sorts of publicity problems. A very Crunch Time episode.

The squeakiest wheel

In labelling the show 'The Hubbert Peak', *The West Wing*'s writers were tipping their hats to M. King Hubbert, a former Shell employee and the father of what has become known as the peak oil movement. The Peak is the summit of a graph Hubbert came up with to predict what would happen to the world's supply of oil as the last century progressed. The graph looks like a narrow bell, and the pointy bit at the top is the Hubbert Peak.

Its shape is determined by how resources are exploited: oil is discovered, but then new discoveries tail off, production reaches maximum, and then oil runs out. Hubbert did the sums back in 1956. He predicted American oil production would peak in 1970, and world oil production in 1995. The peak of oil production is when oil wells simply can't gush any faster – after which it slows down to a trickle as far as supply is concerned.

According to the gloomy prognostications of the most radical of Hubbert's disciples, from that point onwards, anyone who wants more energy is going to have to take oil out of the mouths of someone else, either by paying more for it, or physically locking up supplies – with military force if necessary. If oil were to peak quickly and sharply, with oil wells running dry much faster than they are being opened up, the world would be an unhappy place, they say. Britain's former environment minister Michael Meacher is a born-again peakster. He says: 'When the oil runs out the economic and social dislocation will be unprecedented.'

Why will peak oil time be Crunch Time? Take China, for instance. Experts reckon China needs to keep growing at about 8 per cent a year to avoid huge social ructions among its enormous population. That means it will need three times the amount of oil it consumed in 2005 in about twelve years from then. If world oil production has peaked, then the Chinese will have to pay more for their oil than the next guy. Or, say the scaremongers, they could just take it.

When you tune in to the financial news and you hear that Brent crude is up, down or in-between, you're hearing that it's a commodity – one barrel of oil is much the same as the next, sort of, and whoever pays most for it on the open market gets it. But there's a catch. Nearly 80 per cent of oil production is controlled by state-owned companies whose governments can act to frustrate the market. Governments can cream off cash and leave oil fields with ageing infrastructure and drilling equipment. They can do barter deals to take oil out of the market altogether. Or they can withhold supply to put the squeeze on customers, or gush it out to undermine alternative energy investments.

Are we there yet?

The sixty-four trillion dollar question is of course – how much oil is left? The answer is measured in barrels – the old wooden, metal-hooped things they used to ship the stuff in when they first brought it out of the ground in Titusville, Pennsylvania back in 1859. A barrel holds 159 litres and the world's current annual consumption is around 30 billion barrels, or around 0.03 trillion barrels a year.

And where is the peak? Have we hit it yet? The oil industry itself reckons that when commercial exploitation began there were between 6 and 8 trillion barrels of oil on the planet, under the ground, in the rock and under the oceans. Experts reason it should be possible to extract half of this, with the rest being simply too difficult to get to. So we can count on 3–4 trillion barrels. The big oil companies estimate that we've taken about a trillion barrels already. At present levels of consumption – and deducting what has already been consumed – that gives us enough oil to last between 66 and 99 years. The 'oiliest' of the big oil companies, ExxonMobil, reckons the peak is a long way off, and echoes this in one of its ads: 'Oil is a finite resource, but because it is so incredibly large, a peak will not occur this year, next year or for decades to come.' The glummest industry assessment comes from BP, which thinks the global oil peak may come as early as 2015.

Peaksters reckon we're pretty much there already, with the peak coming sometime before 2010. The United States Congress now has its own Peak Oil Caucus. According to Michael Meacher:

> Already four-fifths of the world's oil supply comes from fields discovered before 1970 and even finding a field as large as Ghawar in Saudi Arabia, the world's largest, would only meet global demand for another 10 years ... Almost all expert opinion agrees that [peak oil] is fast approaching, possibly within five years, almost certainly within 15 ...

Peaksters reckon that a tightening market for oil will mean that our lifestyles will be destroyed. The way we produce food, get from place to place, entertain ourselves, and manage our lives will have a head-on collision with reality – just like Josh Lyman. The peaksters have us heading back to the Stone Age. Some are so gloomy they think we've used up the planet's best energy resources for civilisation – forever. Their websites carry titles like 'die-off' and 'wolfatthedoor'. Their books have titles like *The Coming Economic Collapse*, *End of Suburbia: Oil Depletion and the Collapse of the American Dream*, and of course the classic *Gardening When It Counts: Growing Food in Hard Times*. These are the peaksters that oil lobbyists really like – alarmists with attitude.

But the loonies in the hill-top cabins with their guns and dogs (and former UK government ministers) aren't the only ones who see hard times up ahead. The government of Sweden does too. In 2006, it published a report from the Commission on Oil Independence, chaired by the country's then prime minister, explaining exactly how and why they were planning to make Sweden oil-free by 2020. The country's goals are to eliminate the use of oil as a fossil fuel in heating, and cut its use by half in transport and by a third in industry.

But it's not peak oil that is Sweden's primary motivator. Even though Sweden sees oil running out, the country thinks the real worry isn't that oil will run out ten, twenty, or 30 years from now, but that it won't run out at all. Oil may very well remain so abundant, and cheap, that there will be too little incentive to limit its use and develop alternatives. The real worry, for the Swedes (and for the rest of us) are the pollution and climate consequences of carbon-based forms of energy. Let's take a quick trip to China to explain.

Dirty old Inner Mongolia

In 2006, the governor of a province on China's Mongolian border presided over one of the most phenomenal rates of economic growth anywhere in the world – nearly double China's own impressive national average. His reward? The party bosses in the central government made him write a humiliating letter of abject self-criticism. The governor's problem was not the growth, but how it came about.

It wasn't an Enron-style accounting trick that boosted Inner Mongolia's economy. It was unauthorised construction. And it only came to light through the kind of 'bad luck' that plagues China's building industry – an industrial accident that killed several workers. The accident, at a place called Xinfeng, got nationwide media coverage, not for the accidents, but for what was being built there – a coal-fired power station, costing more than £200 million. A £200 million power plant without planning permission that no one at party headquarters knew about.

On closer inspection the Beijing authorities found it wasn't just one power station. Inner Mongolia was building a dozen unauthorised coal-fired power stations. The province was building dirty generators with a total generating capacity of 8.6 gigawatts, the equivalent of fourteen Xinfengs, at a cost approaching £3 billion. By way of comparison, Europe's largest coal-burning power station, Drax in North Yorkshire, generates 3.7 gigawatts of electricity and pumps out 4 per cent of the UK's carbon dioxide. In just one of China's provinces, the equivalent of over 9 per cent of the UK's current carbon emissions was under construction – without the government's knowledge in the world's largest 'planned' economy. Global warming, here we come. Acid rain, too, which has been pouring down across a third of the country, devastating homes and farmland.

China's cabinet, the State Council, decided to make an example of Governor Yang Jing, along with his deputies and officials, hence the letter. But lenders, building contractors and officials in Inner Mongolia and across China are betting that it's unlikely that £3 billion pounds of investment in energy infrastructure in the world's most energy-hungry country will be ploughed back into the earth. If they build it, they are thinking, they will use it.

Inner Mongolia isn't worried about Hubbert's Peak. Its problem isn't that coal will run out, but that there's simply too much of it. There are more than 1,300 coal mines in the province, and the coal that comes out of them is used for generating electricity and for producing steel and aluminium – needed to build houses and cars for the province's people, and to earn export income. So Inner Mongolia's conundrum is a real Chinese puzzle, so to speak: on the one hand, it needs to grow to feed and house its population, but that growth brings pollution. That pollution isn't costed into the price of a chunk of coal dug out of the ground by a miner, so coal is literally dirt cheap.

Although China's big power companies have been ordered to ensure 15 per cent of their electricity is generated from renewable sources by 2020, the cheap price of coal makes life tough for alternatives. The Chinese are trying to develop new forms of energy. Inner Mongolia has a wind farm at Huitengxile, pumping out a twelfth of the electricity generated by a coal station like Xinfeng. But China's nasty, dangerous, low-cost mining means electricity from the wind farm is twice as expensive as power generated by coal.

Now the Chinese government is starting to act over environmental concerns. Yet how effective can that central power be in a country with 30 or so provinces, each one big enough to put up a dozen or more power stations without anyone really noticing? In 2005, China brought online 68 gigawatts of power. In 2006 this became 80 gigawatts, pretty much the entire UK supply that Britain has built up over a century. So far, most of that power has come from coal.

It's not all bleak, though. To generate £1 million in economic output, China needs eight times more oil – or its energy equivalent – than Japan does, so there's huge capacity to reduce energy use through improving efficiency.

Nonetheless, within the next decade, China will become the world leader in greenhouse gas emissions.

Alarmingly, China isn't the world's coal colossus. Although it uses 40 per cent of the coal dug out of the earth every year, it only has 12 per cent of the world's recoverable coal reserves. The Saudi Arabia of coal is none other than the United States of America. Saudi Arabia has more than a fifth of global oil reserves. America tops that: it has more than a quarter of the world's recoverable coal – about 270 billion tons, spread out from Wyoming to Pennsylvania, and for America it's becoming the power source of choice for generating electricity. Hello again, global warming.

Glimmers of hope

All's not lost. The Technologists may yet swoop in to save us. Although burning coal produces lots of nasty side effects, the main problem that affects climate change is one of chemistry. Every ton of coal burned produces two tons of carbon dioxide; this ends up in the atmosphere, or the oceans, where it starts the process of turning it into a diluted form of carbonic acid. There are good reasons why you don't keep goldfish in fizzy mineral water.

Cleaning coal up is an expensive business, but the big problem is what to do with the extra carbon dioxide. Researchers think they've found a way. In 2004, scientists at Princeton published a research paper with the snappy title 'Transportation Fuel From Coal With Low CO_2 Emissions'. Picking up on this, the Chinese government launched a pilot scheme in the vast reaches of Inner Mongolia, the first of its kind in the world. In the sandy wastelands of Ordos in the south of the region, a project is underway to turn coal into something called dimethyl ether (DME), a colourless water-soluble gas. DME is non-toxic and environmentally friendly. It can also be used for transport. One problem is that turning a ton of coal into liquid fuel needs three tons of water, so putting it in a desert may not have been a great idea, but that's where the coal lies.

Who knows, the Ordos plant might just start the process by which China cleans up its act. The question is, will we get there before the crunch comes? As with all Crunch Time problems, there are more dimensions to the energy issue than there are rigs in the North Sea. Even if we were able to turn coal into a perfectly clean fuel, there's not enough coal in the world to replace our favourite fuel, oil.

Blood and oil

Unlike the Swedes, the United States government is publicly bullish about energy. In 2006, its Energy Information Administration (EIA) revised upwards its estimate of the amount of oil left in the world. There was so much, said the EIA, that China and India could both grow their economies on the back of the stuff sitting under the Middle East for at least another quarter of a century.

Others are less certain, among them a hawkish Republican Senator for Indiana, Dick Lugar. Lugar is a sixth-termer in the Senate, one of its oldest and wisest heads, and he thinks America is suffering from a structural inability to take a decision. He reckons the competing interests of oil companies, car companies, environmentalists, truckers, farmers, consumers and government agencies act to cancel out initiatives or compromises that serve the broader public interest.

He sees six threats to energy security, none of them related to the idea of peak oil or oil supplies running out:

1. **Natural disasters, war and terrorism** – Hurricanes in the Gulf of Mexico, the conflict in Iraq, unrest in Nigeria, sabotage. All threaten supplies.

2. **Increasing competition** – China and India want more energy to fuel their economies, raising prices and creating the potential for conflict.

3. **Supplier power** – In a world of limited supply, the 'haves' can start to flex their muscles. Venezuela and Iran have threatened to use oil as a diplomatic weapon. Russia has used the gas supply to browbeat the Ukraine.

4. **Propping up corruption** – Imports normally generate growth, but the cash paid out for oil does not generate as much growth in producing countries, and bankrolls authoritarianism and corruption.

5. **Climate change** – Inefficiency and pollution have plagued our use of fossil fuels.

6. **The development cost** – Rising energy prices threaten the poorest most.

These conditions might be negotiable in the short and medium terms if oil fields were run by socially responsible, politically secure producers who maximised production when demand went up. Except they aren't. Three-quarters of the world's oil is parked right under the Middle East. And the vast majority of oil fields are afflicted by at least one of three problems: lack of investment, political manipulation, or the threat of instability and terrorism.

As recently as 2002, spare production capacity exceeded global oil consumption by about 10 per cent. As world demand for oil has rapidly grown, spare capacity has declined to less than 2 per cent within four years. With supply that tight, even minor disruptions of oil can drive up prices.

One conservative estimate puts the United States' oil-dedicated military expenditures in the Middle East at US$50 billion per year. But clearly even that gargantuan effort can't guarantee energy security. Even George W. Bush, oilman, admits America is too hungry for the black stuff. In 2006, the United States spent as much on oil imports as on three years of fighting and rebuilding in Iraq.

Rising energy prices, news reports of hostile oil producers, and the energy shocks experienced after the Katrina and Rita hurricanes, and petrol prices heading up towards US$3 a gallon, woke Americans to their energy vulnerability.

Cold turkey

The United States has the world's most serious oil addiction. As Alan Greenspan, a former Federal Reserve chairman, explained to nervous senators on the Foreign Relations Committee, almost one out of every seven barrels of oil produced in the world is burned on America's roads. Motor manufacturers and the great car-buying public aren't exactly helping. In 1987, the average light duty vehicle travelled 22.1 miles on a gallon of petrol. Nineteen years later, that figure is down to just 21 miles. Engine technology hasn't stood still in that time, but US fuel efficiency standards have. Today's family cars have the performance of a seventies supercar, while US fuel efficiency standards are amongt the lowest in the world – much lower even than China's.

Despite all that is to be said against big cars from a Crunch Time perspective, American car manufacturers still put their faith in pick-ups and SUVs. In 2006, at the start of what the peaksters are saying is the next long oil shock, General Motors offered its customers a gasoline subsidy, capping the price of a gallon at $1.99 for one year for buyers of Hummers and other gas guzzlers.

Meanwhile, US legislators are putting their feet on the energy accelerator. America's 55-mph speed limits – brought in to cope with the oil shocks of the early 1970s – are being raised. Texas upped speed limits to 80 mph on some roads. And all of this when a great deal could be achieved without consumers feeling much pain at all.

Cars, SUVs, vans and pick-up trucks account for over 43 per cent of America's consumption, the biggest single use. Air travel accounts for 6 per cent. According to their Department of Transportation, two-thirds of car journeys and half of all plane rides across America are 'discretionary' – that's how bureaucrats say 'unnecessary'. When oil prices start hitting American pockets big time, there's somewhere to go – not for a drive to the airport. Or for that matter the corner shop. Eliminating discretionary journeys could cut America's oil consumption by 31 per cent. Americans and the rest of us used to suburban living might just have to start talking to their neighbours while car pooling or riding the bus to the mall.

Petrol prices are beginning to have some effect on the cars Americans buy. Ford's SUV sales slumped by nearly 30 per cent in 2006. But even though the price of petrol is rising in America, its drivers are still buying it in record quantities. Americans bought 10 per cent more gasoline in the first half of 2006 than they did in the first half of 2000, even though the price was 75 per cent higher.

Glimmers and flickers

Progress on energy efficiency is also appearing in the investment world. The entrepreneurial vanguard that brought us the internet and transformed telecommunications is turning its attention to alternative energy. American venture capital targeted at alternative energy projects more than tripled to US$315 million in the first half of 2006 compared to the first half of 2005. It's a low base compared to the multi-billion investment budgets of the oil industry – but it's growing exponentially.

In late 2006, Virgin billionaire Sir Richard Branson pledged ten years of profits from his transport empire – trains, planes and automobiles – to alternative energy investments – a sum of over $3 billion. Critics were quick to ask if the money shouldn't go into greater fuel efficiency for the transport business. But alternative energy is no longer just a niche area for environmental idealists and companies trying to improve their public image. If the ravenous dogs of capitalism wake up to the emerging economics of Crunch Time, there's no telling what feats of human ingenuity we may see.

Admitting there's a problem is a necessary first step to solving our energy issues, but it doesn't mean that those problems will be solved. It won't matter that when disaster strikes, we have a sense of realism about our vulnerability. It won't matter that we were thinking of cycling to work, or heating our water with solar panels. And we're still a long way off. Even in California, where voters tend to be environmentally sensitive and where pollution provides a strong extra impetus to cut petrol use, business interests have succeeded in discouraging alternative fuels and transportation technologies. Since 1979, California's legislators have tried a variety of approaches, only to be frustrated by the oil industry and motor manufacturers. A proposal there to cut oil use 15 per cent by 2020 is supported by 'Governator' Arnold Schwarzenegger, but opposed by Big Oil. Still the average Californian burns through more than 400 gallons of gas a year.

Politicians don't mind talking green, but even brave ones shudder when they recall the epic unpopularity of Jimmy Carter's energy programme in the 1970s. Carter's unheeded calls for energy sacrifice remain a cautionary example for many office holders, leader writers and political strategists. Conventional political wisdom holds that voters will punish anyone who forces significant energy sacrifices on them. This is a major oversimplification, but it's true that voters are not eager to pay higher prices for energy, wait in line at petrol stations, or see their freedom to drive curtailed. A 2006 poll asked about 1,500 Americans which of five options was 'the best way to reduce U.S. reliance on foreign oil'. Two per cent chose increasing the gas tax. Building new nuclear plants or enforcing stricter mileage standards fared little better at 6 and 8 per cent respectively. Respondents gravitated toward general trends that were unlikely to affect them personally, with 52 per cent endorsing increased government investments in alternative energy sources and 20 per cent choosing to relax environmental standards for oil and gas drilling.

Breaking through a political logjam often requires a crisis that focuses our collective psyche in a way that achieves a consensus. But consider that the combination of 9/11, the war in Iraq, the conflict on the Israel–Lebanese border, the nuclear stand-offs with Iran and North Korea, the Katrina and Rita hurricanes, sustained a gasoline price of $3 per gallon, and several other severe problems have not created a consensus on energy policy.

The question is, what kind of a crunch will have to happen before public opinion and our political structures are sufficiently energised to actually shift things radically away from fossil fuels? Perhaps something out of the pages of the peak oilers' textbooks: oil over $100 a barrel and a worldwide economic downturn. The point is one that *The West Wing*'s clever writers know only too well – we're all like Josh Lyman, attracted by the gleaming and powerful Humvee in the corner. At what point does getting behind the wheel for a test drive become simply too expensive?

CHAPTER 6

Big Bad Business

The myths and realities of corporate power

> *Our merchants and master-manufacturers complain much of the bad effects of high wages in raising the price, and thereby lessening the sale of their goods both at home and abroad. They say nothing concerning the bad effects of high profits. They are silent with regard to the pernicious effects of their own gains.*
>
> Adam Smith, *The Wealth of Nations*

Some things Mike and Adrian agree on. One of those things is that, like it or not, living an ordinary life in the 21st century means being part of something very, very big. Take the following apparently insignificant vignettes from our plainly ordinary lives.

A random Tuesday in September, 5 pm. Mike is driving home from a meeting across Sydney, stops at a service station to fill up with petrol and buy a bottle of mineral water. Arriving home, seven-year-old Max is in front of the television, playing PlayStation® and six-year-old Joel is in the backyard playing with his yellow truck. Claire is sitting at the table reading a textbook.

Over in London, the day is just starting. Breakfast for Adrian is toast, coffee and a Nutri-Grain Bar in the car, on the way to the train station. Meanwhile, little Ella has turned on Cartoon Network before school and Linda is scanning through *The Times* online.

Innocent enough routines, probably repeated with hundreds of millions, or even billions of minor variations across the globe every day. Without even stopping to think, almost every moment of every day, the intricate web of transactions that make up global business are subtly reinforced and strengthened. By us.

How so?

Toyota, which made Mike's car, had annual revenues in 2005 of US$185 billion. This sits between, say, the 2005 GDP of Portugal ($183 billion) and Finland ($193 billion). The petrol was sold by BP, a minuscule fraction of its $277 billion total sales in 2005, which means it trails slightly behind the GDP of Saudi Arabia, at $308 billion. Neverfail, the manufacturer of the mineral water casually purchased at the BP service station, is part of Coca-Cola Amatil, a $3 billion spin-off from the Atlanta-based Coca-Cola Company (2005 revenues of $23 billion).

Max, the PlayStation® zombie, is an early capture for Sony, a company that raked in $66 billion in revenues in 2005, with over 160,000 people on the payroll. That's a bit higher than the GDP of Bangladesh at $61 billion, but then Bangladesh contains more than 147 million people. Even little Joel's Tonka truck is produced by a global toy behemoth Hasbro, which pulls in revenues of nearly $3 billion, and Claire's textbook is produced by Penguin, owned by the world's biggest educational publisher, the $7 billion British corporation Pearson.

Over in London, the portrait is equally easy to paint – not even Adrian's toast is free from the taint of global capitalism. The bread, purchased from the local Tesco supermarket (2005 revenues of $71 billion), was baked by Hovis, part of RHM, valued at getting on for $2 billion.

Coffee by Lavazza, an Italian company with a group turnover close to $1 billion, just from coffee; Nutri-Grain Bar by Kellogg's (2005 revenues of $10 billion); car by Volkswagen AG (2005 revenues of $123 billion, equal to Israel's GDP); rail trip to London by Southeastern trains (owned by a joint venture between transport players the $3 billion Go-Ahead Group and Keolis at $2 billion). Ella's entertainment is courtesy of American showbiz octopus Time Warner, with 2005 revenues of $44 billion, while Linda's *The Times* feeds Rupert Murdoch's $24 billion beast, News Corporation.

Big business, it seems, is everywhere. It sits at the breakfast table, drives with us to school, infiltrates our clothes and inhabits our homes and workplaces. It informs us and keeps us ignorant, it encourages our bad habits and profits from their treatment. It's with us at our lying down and at our waking. Like Charlie Chaplin in *Modern Times* (filmed in the middle of the Great Depression), we are but cogs in the wheels of the great machinery of the global corporation. Stamped 'customer' at birth, the brand remains till we die. And, short of stranding ourselves on a desert island, it seems there's little we can do about it.

This, Mike and Adrian can agree on. But we don't agree about what this means for how we should live our lives in Crunch Time.

Corporate power – a lesson from history

Trade cannot be maintained without war or war without trade.

Jan Pieterszoon Coen, 1614, on how to do business in India

Here's what Adrian reckons. The fact that enormous global corporations dominate our lives is neither novel, nor especially scary. Today's big business is altogether less powerful and less threatening than it has been for hundreds of years. To prove the point, look at a modern landmark of the City of London, one of the world's great financial centres – One Lime Street.

Today it's home to a complex and twisted mountain of inverted stainless steel pipes and fittings, an award-winning architectural marvel built by Richard Rogers, into which each day hundreds of insurance brokers disappear, toting their leather folders and laptops. Here, the risks of global capitalism are underwritten by the syndicates that call this building, and the insurance market that sits within it, Lloyd's of London, home.

But, impressive as it is, the kind of 21st-century business that flows through One Lime Street today pales into insignificance compared to the enterprises that were planned and executed on this spot as recently as 150 years ago.

Then, the site was headquarters to the most significant global corporation the world has ever seen, or is ever likely to see, and the directors behind their leather writing desks exercised the kind of power that would make the chiefs of BP, Toyota, News Corporation and the rest green with envy.

At the height of its power, the United Company of Merchants of England Trading to the East Indies – better known as the East India Company – had an army of 200,000 and really did rule the world. Its authority extended across what today is India, Pakistan, Bangladesh, Burma, Singapore and Hong Kong. A fifth of the world's population was under its power. At various stages in its 274-year history, under the auspices of trade, it militarily defeated Imperial China, occupied the Philippines, and conquered Java.

The Company's history began with a Royal Charter – an enforceable monopoly from Queen Elizabeth I on 31 December 1600. Initially a speculative venture to trade pepper and spices from Indonesia, the Company became seriously rich by conquering India in the 18th century, laying the foundations of the British Empire.

In the 1750s, three decades before the First Fleet set sail to Australia, and two decades before disgruntled American colonists threw their (East India Company) tea chests into Boston harbour, Robert Clive, the Company's most brilliant servant, engineered a local coup in Bengal. He publicised it as the Battle of Plassey, after a somewhat dubious military encounter, and ended up running the place. Administering an entire country required armed force, and armed force required taxes. Walloping the Bengalis militarily, the East India Company won itself the right to tax over 20 million people. Revenues of some £2–3 million a year – the equivalent today of Switzerland's GDP – were shipped back to London, to the delight of the Company's directors. A dozen or so years after the Company took over running Bengal there were scores of rich Englishmen. And, owing to an 'unfortunate' famine shortly after, millions of dead Bengalis.

The history of the Honourable Company includes swashbuckling stories that would make the hair stand up on the back of the necks of the most swaggering Crunch Time corporate execs. These were the corporate gunslingers who hired Captain Kidd to undertake pirate raids in the South Seas, attacking any ship that sailed under the flag of Britain's enemies. These were the wheeler-dealers who leased St Helena to the British government as an upmarket Guantánamo Bay for Napoleon to see out his final days. These were the repressive rulers who provoked the Indian uprising of 1857.

This was the era before 'corporate social responsibility', and the Company's executives pioneered the kinds of business practices that get modern multinationals pilloried and their managements (occasionally) jailed.

Today's fat cats, insider traders and dodgy dealers are as nothing in comparison. The East India Company enriched its corporate elite to a degree barely comprehensible by today's standards. After the 32-year-old Robert Clive installed a compliant local to help him run Bengal, the entire contents of its treasury were 'offered' to him, making him unfathomably rich. At a public inquiry into the probity of the affair, Clive's defence was simple: he could have had the lot, but only took most of it.

More than the massive salaries and dubious methods, the Company pioneered multinational exploitation of the global 'South', as the poor world has become known (much to the chagrin of Australians and New Zealanders).

Back in the 18th century, fashionable Europe craved silks, spices, porcelain, cottons and a decadent new drink: tea. These came from India and China, but the Indians and Chinese didn't crave Europe's woolly undergarments and other manufactured delights. So the Easterners made a lot of money. Gold, and especially silver bullion, flowed from West to East.

The East India Company forcibly stopped this transfer of precious metal when it started helping itself to chunks of India. It began using the country's own wealth to pay for the exports of its goods back to Britain. Many Indians reckon that the East India Company's way of doing business was the earliest cause of their country's slide into poverty. When the Company arrived in southern Asia, India was a relatively rich nation with sophisticated industries of its own. Britain plundered it, and used the money it found there to fuel its own industrial expansion. A little later on, the Company used Indian-grown opium to make millions out of exporting addiction to China.

When the Company's charter finally expired in 1874, the London *Times* noted: '[I]t accomplished a work such as in the whole history of the human race no other company ever attempted and as such is ever likely to attempt in the years to come.'

Rupert Murdoch, eat your heart out. In all, says Adrian, Crunch Time capitalism is much more caring and sharing than in years gone by.

Corporate power – a lesson from today

> *Corporations have been enthroned and an era of corruption in high places will follow, and the money power of the country will endeavour to prolong its reign by working upon the prejudices of the people until all wealth is aggregated in a few hands and the Republic is destroyed.*
>
> Abraham Lincoln, 1864

Still, as we saw at the beginning of this chapter, there's no doubt that we live in an age of corporate behemoths.

While the East India Company was indeed the daddy of them all, it did rule at a time when there weren't that many other corporations spanning the globe, bending its resources to their own ends. Today, there are over a thousand companies with revenues in excess of a billion American dollars.

Mike picked up the story in a phone call one random Tuesday evening in September.

'It's not about cinnamon and chintz anymore. It's not even about planting flags in far off parts of the world anymore. It's much more insidious than that.'

Mike reckons that in Crunch Time, corporations are complicit in perhaps the greatest conspiracy ever: the perpetuation of Western, consumerist values across the globe; the systematic plundering of the earth's resources for trinkets: designer baby clothes, mobile phone handsets, consumer electronics, lifestyle excess, luxury travel, the whole 'too many toys' thing we saw in Chapter 2 ... the list goes on. Corporations, their marketers and advertisers, take advantage of our everyday human desires and insecurities to sell us a bunch of stuff we don't need, to perpetuate a wasteful system that – as we have seen in earlier chapters – the rest of the world simply can't ever afford.

From Mike's bookshelf – admittedly a biased selection – just three of the books whose hundreds of pages outline the charges against the 21st-century multinationals:

1. *No Logo* – Canadian journalist Naomi Klein's account of the not-so-slow progress of corporate branding into every part of modern life: our cities and landscapes, our youth, our education systems; the way brands are used to prise big profits out of customers whether they can afford it or not; and the anti-corporate movement's protests and boycotts against all this.

2. *The Silent Takeover* – UK academic Noreena Hertz argues that the declining power of the state and the increasing power of business over the past 30 years is undermining democracy and leaving us at the mercy of the market.

3. *One Market Under God* – Social critic Thomas Frank claims that Western society has elevated the market and its non-values above everything else.

Want more? Type in 'anti-corporate' on Amazon.com and well over a thousand titles come up, with catchy titles like *The I Hate Corporate America Reader: How Big Companies from McDonald's to Microsoft are Destroying Our Way of Life*, and the not-so-catchy *There Is An Alternative: Subsistence and Worldwide Resistance to Corporate Globalization.*

'Naomi Klein?' replied Adrian. 'What better example of a global brand is there than Naomi Klein? I'm surprised she hasn't started putting No Logo on an exclusive line of clothes distributed through Wal-Mart.'

Cynicism aside though, Mike insisted that the abuse of corporate power wasn't consigned to history. The crimes levelled at the joint stock company are many. Apart from the 'conspiracy to promote consumerism' charge, which we'll look at later, here are the big ones.

1. *Exploitation: the sweatshop issue*

You're the production manager of a children's clothing company based in the United States, and have been asked to recommend where to locate a new factory. You have been told to base your decision on cost and productivity.

How about the home of the free? Well, in the United States there's no national minimum wage, and factory workers make around $7 an hour. The UK has a minimum wage of £5.05 (equivalent to $9.85). The Germans don't have a minimum wage but they're highly unionised with big government-mandated social costs passed on to employers.

So should the factory stay in Idaho? Take a look at the hourly wages for factory workers in 'developing' countries below.

Country	Hourly pay (US$)
Mexico	0.50
Honduras	0.43
China	0.28
Nicaragua	0.23
Indonesia	0.20
Burma	0.04

Indonesia looks a little more attractive than Idaho. Even with the security situation. Why not subcontract your manufacturing to a developing nation where workers will spend their lives producing your goods at a fraction of the cost of the folks back home? And without demanding any of the pesky limitations on hours worked, let alone demands for pensions, holiday and sick pay, or health and safety provisions.

In fact, while we're at it, why don't we close down our existing factories and move them to more cost effective locations too?

Making things (T-shirts or software) where the cost of labour is lower certainly makes sense for management (if not the newly unemployed staff in shutdown factories). It means cheaper goods, higher profits (higher bonuses) and it means people who would be otherwise poor and unemployed in developing countries now have jobs.

And let's face it, corporations didn't make countries

poor (well, maybe they did a couple of centuries ago, but that was then and this is now).

The problems come with the regulation of jobs in developing countries.

Workers and employers aren't paying for the kind of welfare and health-care provision Western countries demand. Conditions aren't regulated, regulations aren't enforced, and enforcement isn't acted upon. Stories of fourteen-hour shifts, seven-day working weeks, child workers, bonded labour and the worst kind of Victorian workhouse conditions are not difficult to unearth.

And the problems don't end there. By choosing to pay low wages (either directly or at arm's length through subcontracting) and operate in countries where standards are lower, companies are complicit in the not-so-slow transfer of wealth from the poor world to the rich that we have seen in earlier chapters. While sweatshop workers put in the hours, the rat race's big winners in New York and London and the rest of the rich world make off with the cheese, and that includes us.

And that's what many of the world's best-known brands have been caught doing: producing clothes and other goods at a tiny cost in some crummy Third-World hellhole, and then packaging, advertising and retailing them at designer prices in elegantly furnished stores.

It's a very Crunch Time kind of crime. But it's joined in the dock by a much older sort of crime, and one that's not going away.

2. Corruption: trousering the profits

As we will see in Chapter 9, the corrupt corporate sector also dominates national debates through the obscene amounts of money it provides to political parties and lobbyists. Corrupt, clearly.

But, corporate corruption comes in all shapes and sizes. Here's a little example from capitalism's capital, Manhattan. Jack Grubman, a $20-million-a-year telecoms analyst with the world's largest bank, Citigroup, decides he must get his two little daughters into a plush Upper East Side nursery, the 92nd Street Y. He approaches his boss, Citigroup CEO Sandy Weill, the self-declared king of Wall Street, and asks if Sandy might be able to swing it for the Grubman twins to get into kindergarten.

The boss is happy to help. Calls are made. Citigroup pledges a million bucks to the nursery school. Places are found. Does Sandy want anything for keeping one of his top analysts sweet? Well, aside from the day job, Sandy is also a director of AT&T, and kind-hearted Jack apparently agrees to recommend AT&T stock to his investors. A favour is returned, the only victims – the people who bought the stock on Grubman's say-so (and perhaps the parents whose childcare places have suddenly evaporated).

But corruption goes beyond the snobbish concerns of Wall Street's superstars. Take French oil company, Elf. In November 2003, the ex-president of Elf, Loïk Le Floch-Prigent, was sentenced to five years in jail for overt bribery and secret political party funding, after confirming an open secret – that for decades the state-owned energy giant provided a 'black box' into which funds were secreted to cover all manner of political shenanigans.

General De Gaulle created the black box system so that Elf could grease its way into oil contracts and challenge its British and American rivals away from the gaze of French tax inspectors.

But guess what? Putting a slush fund intended for bribery at the disposal of executives corrupt enough to engage in such practices was like passing round a vodka bottle at an AA meeting. The weak and greedy (who in this case happened also to be the rich and powerful) just helped themselves.

After eight years of investigations, the trial opened in March 2003, with 37 defendants and 80 lawyers crowded into a Paris courtroom. Revelations were rife. In the space of four years, between 1989 and 1993 (the year before Elf was privatised), senior Elf executives skimmed hundreds of thousands of pounds from Elf's black box of secret funds straight into their personal accounts, often with the approval of then President, François Mitterrand. Le Floch-Prigent was found, among other things, to have used Elf money to buy a multi-million-pound apartment in Paris and a chateau near the capital, and to pay off his ex-wife, Fatima Belaid – also in the dock – in a hefty divorce settlement.

During Le Floch-Prigent's tenure, the amount of money sloshing through the secret accounts, known as 'the kitchen', multiplied tenfold, with a couple of million pounds a year allotted just for paying off French politicians – names mentioned included the President's.

In Australia, too, the tendency for people to behave differently at work than they would at home is often on display. For example, in 2003, the country's monopoly wheat export company, the Australian Wheat Board, was paying millions of dollars in bribes to the late Saddam Hussein through a third-party trucking company in exchange for wheat contracts, just as Australia was preparing to invade Iraq to depose the dictator.

Everywhere they operate, corporates have a tendency to seek political favours, allied with the cash to buy them. They bend and break the rules, they push the boundaries, and they do it for the personal gain of the powerful people in the company. It's occasionally blatant, always repugnant.

3. Unaccountability: getting away with it

By dint of their size, power and wealth, global corporates are arguably the single most important influence on the global economy today. But corporations aren't accountable to the people over whose lives they hold sway. They might produce accounts, but as the bosses of Enron noted, getting them endorsed merely requires the right kind of accountant. Clearly not too difficult to find.

We don't vote for the management of corporations. The rules they follow, where there are any, are determined in shady meetings in five-star hotels, under the auspices of bodies that are above the states we live in: the World Trade Organization, or the Council of Foreign Relations, or the Trilateral Commission (a body of over 300 senior people which meets every three years to thrash out the rules of globalisation), or the World Bank. The rules that these bodies (unanswerable to anyone) decide upon are the rules by which the game is played – whether we like it or not.

Businessmen rail against any restrictions. The chairman of Dow Chemical once dreamed publicly 'of buying an island owned by no nation and of establishing the world headquarters of the Dow company on the truly neutral ground of such an island, beholden to no nation or society'. Multinationals threaten to up-sticks and go elsewhere whenever governments have the temerity to threaten to raise taxes or toughen safety standards.

Corporations are supposed to be accountable to the law, to government and to their shareholders. But they spend money on lobbying to weaken or change laws, they spend money on politicians (see the example of Elf), and, if they keep paying dividends, their shareholders aren't likely to ask too many questions either. The 'government relations' and lobbying power of corporations isn't matched by anything on the other side of the political debate, so inevitably big business gets a better hearing than little old you or me.

By this argument, those who are the victims of pro-business policies and decisions, including sweatshop workers, local farmers and businessmen who are bankrupted and the global 'South', are ground underfoot by the remorseless march of heartless global capitalism.

So there you have Mike's shortlist of allegations against the corporate world – exploitation, corruption and unaccountability. But by the end of the phone call he'd thrown in the rest. The world is becoming blander, a Starbucks and McDonald's on every corner. Our ecosystems are being destroyed as global corporations reach into the furthest corners of the earth and rape the ground, sea and air, only to return home to count the profits. The anti-logic of corporate operations has turned the world into a 'race-to-the-bottom' as countries and regions within them compete to provide the lowest standards and most attractive environment for the profiteers. They don't pay tax. They're amoral, dirty, dehumanising, aggressive and generally a bad thing.

The case for the defence

It's never good to interrupt someone in full flow, and Adrian decided not to point out that it's a bit rich for us to sit here in centrally heated or air-conditioned comfort, travelling slowly but comfortably in our privately owned cars (fuel courtesy of the heinous and bloodthirsty oil industry), entertaining our kids with the poison fruit of Sony and Time Warner, and when the contradictions end up giving us a headache, popping a GlaxoSmithKline pill to make us feel a bit better, then bitch about the terrible influence of global corporations.

Without the innovation of the joint stock company and its ability to risk large amounts of capital on audacious projects we'd still be living in the Dark Ages. Look around. Everything we have and everything we do is pretty much owed to the corporation.

'Mike', he eventually said, 'you're way off track'.

Companies have their faults, for sure. Bosses, being people, often get weird ideas about their place in the world and go off the rails. But when they do, they get caught, and that simply prompts society to figure out tougher ways to police executives who've diverged too far from what most of us think is acceptable business practice. United States Treasury Secretary Paul O'Neill memorably described Enron's $100-billion failure as a 'triumph of American capitalism'. Whether the shareholders and workers whose wealth went down the plughole agree is a different matter: what he was *trying* to say was that it all comes out in the wash in the end.

What about the argument that the power of corporations is growing? The most frequently cited statistics purport to show that companies are now bigger than most countries. For instance, Sarah Anderson and John Cavanagh of America's Institute for Policy Studies published 'Top 200: The Rise of Corporate Power', which reported that:

> Of the 100 largest economies in the world, 51 are corporations; only 49 are countries (based on a comparison of corporate sales and country GDPs). To put this in perspective, General Motors is now bigger than Denmark; DaimlerChrysler is bigger than Poland; Royal Dutch/Shell is bigger than Venezuela; IBM is bigger than Singapore; and Sony is bigger than Pakistan.

Scary stuff. We played around with some of those comparisons at the start of this chapter. Scary, yes, but totally wrong. Comparing a corporation's revenues with a gross domestic product looks good on paper but it's apples and oranges. A country's GDP, with all its faults as a yardstick of progress and well-being, is actually a measure of the total value added of all the products, materials, goods and services produced by that country over the course of the year – in other words, profit. A company's revenues are just the amount of cash that rolled in and out the door that year, the vast bulk of which has to be spent on paying for the cost of producing the goods and services sold, taxes, employee benefits and all sorts of stuff.

So the real size of Toyota by the equivalent of GDP is actually its net profits, closer to $4 billion than the $128 billion figure we quoted earlier. That makes it nowhere near the size of Portugal's economy, and instead about half the size of an economic basket case like Zimbabwe. But unlike Zimbabwe it did produce 6 million cars in 2003, and provided work for 66,000 people.

There are big differences too between the powers of countries, and those of companies. The world's biggest corporation in terms of revenue, Wal-Mart, made profits of just over $8 billion, close to the GDP of Cameroon. Cameroon has nearly 15 million citizens, Wal-Mart has just over a million employees. But the state of Cameroon can do a bunch of stuff that Wal-Mart can't: it can raise an army and a police force, raise taxes, enforce systems, ways of life, and hold political sway over all its people. It can even produce Africa's best football team. Wal-Mart, on the other hand, just sells stuff, and distributes its profits to those who hold its shares. Cameroon's land, people and resources will likely be there long after Wal-Mart is consigned to corporate history.

If an employee, customer or shareholder of Wal-Mart gets cheesed off with the company, they can just go elsewhere: buy their clothes at Kmart, for instance, quit their job or sell their shares. A citizen of Cameroon is a subject of that country, for better or for worse. Emigration is a radical solution for disgruntlement.

And corporations, even the biggest and best, do get binned on a regular basis. Tom Peters, a wacky management guru, wrote a book called *In Search of Excellence* in 1982 in which he examined the characteristics of 48 'excellent' companies. Well the 'ex-' part, at least, was right. Two-thirds of them are now ex-companies.

Yesterday's Wal-Mart, anti-globalisation's whipping boy McDonald's has had to shut down its operations in three countries and is closing nearly 200 branches. Either French cheese-makers storming their restaurants has scared them into global retreat, or consumers are just fed up with eating their burgers. Goodbye global domination, hello commercial reality. And the globalisation of trade and investment means corporations are facing tougher competition from each other too. There are more companies operating in almost every line of business.

OK, replies Mike. Let's take our three accusations one at a time and see how your arguments fare against them.

1. Exploitation

Nobody would say that it's a good thing for workers to be employed in horrible conditions below a living wage. But the fact that companies operate around the world, sell their goods back in the West, and have a Western shareholder base means that they have to have an eye on standards wherever they operate. Shareholders and consumers do care, so the companies have to care.

Part of this is due to Naomi Klein and the rest of the anti-corporate crowd who raise awareness. Look at Nike: pilloried for contracting work to sweatshops, it took the criticism seriously and now imposes a serious code of conduct on its suppliers. The sweatshop problem is solvable in the case of big multinationals. Studies have shown that multinational firms pay more for labour where they operate because they are held to higher benchmarks, and because they are more efficient.

The sweatshop problem is far more of an issue for smaller, private companies that produce goods that aren't readily identifiable as branded – look at that little plastic figure on your key chain or the made-in-China toys that are so cheap down at the supermarket, or those unbranded clothes for sale in the pile-'em-high, sell-'em-cheap supermarket down the road. How can you know the conditions in which they are made? You just can't, because they don't come from a reputable company, or have a reputable brand. It's important to remember that companies that do sell branded goods are held to account for their standards and values.

In any case, without international investment from these companies, many of the jobs they bring would never exist. The challenge, then, is to make sure that they're decent jobs – and Klein and her crowd are doing a good job of piling on the pressure to make sure this happens. As poor countries get richer because of outside investment, standards there will rise as they have all over the developed world.

2. *Corruption*

Crime is crime. Corporate criminals who rip off shareholders or pension funds are no better, and often much worse, than simple bank robbers. Chuck 'em in the clink. If that's not happening, strengthen your laws and sort out your politicians – they're the ones in charge of controlling crime.

And when bad stuff happens, like Enron and the rest, rules get strengthened. Accounting standards and corporate governance is much stronger than it was before all these disasters. Whether it's strong enough remains to be seen, but new laws and executive criminals in handcuffs show the system works, more or less. In developed countries, the regulatory demands on companies are clearly increasing – as they should be – in areas like accounting standards.

More difficult to deal with are the insidious links between politicians and big business – of the Elf kind. But once again, what matters is the robustness of the legal system, and also the corporate culture in which these crimes happen. Culture is down to the people within the companies, which is down to individuals like you and me who work for them.

3. Accountability

Okay, we don't directly vote for company management, but we do vote with our cash. As we saw in Chapter 3, the owners of these big corporations are you and I through our pension funds and investments. Sure, individually we don't have much clout, but together we do. Why do you think companies get so scared when it looks like their reputations are going down the gurgler? Look at Shell, for instance. Greenpeace gave it a hugely hard time when it was going to dump the Brent Spar oil platform into the North Sea and consumers across Europe boycotted Shell products. The company lost something like 20 per cent of its market share in Germany in the month the story was in the news – and in the end the company backed down, turned the platform around, and towed it to a fjord in Norway where it was dismantled at great expense.

All of this is possible only because we live in societies where the activities of companies like this are open to scrutiny. Companies have to publish tons of information on what they are up to – and so they should – and the amount of information they publish is only getting larger. More and more are releasing environmental and social reports as well as financial stuff. Even if you think its total greenwash, it's still more than any companies in non-democratic, non-market-led societies do. What do we know about China's top companies? Even if you're interested and speak Chinese, the information just isn't there. Some management theorists reckon that the future for corporations is 'nakedness': being open, transparent and revealing everything they do all the time, on the web, in the press, to whoever wants to know. Companies that don't do this will suffer.

No, says Adrian. Corporations are just an easy target. They're big. Like all of us, they can do bad stuff. And they undoubtedly command vast swathes of resources. But it's difficult to point the finger at the institution of the corporation per se.

So here we come to a bit of an impasse.

We agree that our lives are dominated by the products of big business. We can even agree that those products are often a good thing – Mike likes that his safe and reliable car transports him around town, that Joel has a nicely produced big yellow truck to play with, and that Penguin's ability to publish high-quality textbooks cheaply allows Claire to study at home rather than spend her time at the library or not study at all. Adrian enjoys his toast and coffee and the rest of the accoutrements of a modern lifestyle, many of which weren't enjoyed by our ancestors who lived before the age of corporate titans.

But we diverge when it comes to thinking about how to curb some of the excesses that inevitably occur when one sector of society comes to command so many of its resources.

For Mike and his Kleinista anti-corporateers, the answer can only be in a radical change in lifestyle, a mass turning away from the fruits of corporate production to a different lifestyle of the kind we described in 'Envirocide'. How might this happen? Through protest, through the kind of movements that were seen in Seattle and Genoa and other anti-globalism riots. Through awareness raising, distribution of information over the internet, through an enormous change in consumer preferences and behaviours, and a steady and determined effort to change both the way we live and the way companies are allowed to interact with society.

For Adrian, the answer isn't so much the magical 'dawning of consciousness' that's hoped for by the new international left. Corporations have power, no doubt, and they use it to their own ends. But surely, this, then, is the secret: align corporations' interests more closely with those of society in general. The ballyhoo protest of Klein and her chums has helped enormously in this: corporations with global brands are now terrified of being caught out doing the wrong thing, and so they should be. If the system has faults, which it does now and will into the future, then surely our efforts should be bent towards plugging those faults, rather than pulling down the whole edifice and all the good things that go with it. Fundamentally, we like our lives. If it ain't totally broke, why try and totally fix it?

There are good, established ways to work to change the system, to limit power where it needs to be limited, and act to influence things to change. We have examined lots of them in this book: in 'Too Many Toys' we saw how we can use our own professional lives and wealth to encourage corporations and markets to take into account our concerns. In 'Envirocide' we saw how changing our own scale of wants and readjusting the way that we live our lives is a crucial first step in solving the broader environmental issues facing the world. The list goes on.

In the end, the answer has to be something in the middle. Mike isn't about to take his family off to a desert island ... he has a hard enough time getting Max to turn off the PlayStation® when it's time to go to school, or avoiding big-brand mineral water at the service station. But that doesn't necessarily make him a bad person. We're all limited by our time and place, and there's a constant tension between what we want to achieve in our lives, our hopes and dreams, and the demands placed on us by our broader concerns for the state of society and the future we're building for the next generation.

CHAPTER 7

Maintaining the Disparity

People, pawns and military power

> *It is not true that the largest civilization is necessarily the winner, nor that the wealthiest always has the upper hand. In fact, a balance of knowledge, cash, military might, cultural achievement, and diplomatic ties allows you to respond to any crisis that occurs, whether it is a barbarian invasion, an aggressive rival, or an upsurge of internal unrest.*
>
> 'User's Guide: Introduction', *Civilization III*

In the civilised world of Sid Meier, players can stamp their mark on the world in a number of ways. They can conquer their enemies through military strength, economic domination, technological and scientific advances, or through cultural domination. Statecraft and diplomacy – which countries to cosy up to, and which to wage war against – are key skills. Still, it's a story of military might that dominates our attention at the beginning of the 21st century, the result of the current state of play in our own macro-civilisation.

These days waging war ought to seem a little old-fashioned. Japan and Germany both reached the commanding heights of the global economy without nuclear weapons and without rearming themselves to pre-Second

World War levels. Both had been defeated and occupied by the United States, and both have since benefited from broad alliance with the USA's post-war global policy aims. Looking at these 'losers' might lead you to think that military force is redundant, that the economic resources required to sustain it just aren't worth the investment.

But you would be wrong. On the great *Civilization* scoreboard there's no doubt that the player currently ahead, the United States, isn't about to abandon its commitment to armed force. In 2005, with just 4.5 per cent of the world's population, it produced nearly a fifth of the world's economic output and consumed about a quarter of the world's oil supply. To keep that position it spends about 40 per cent of global military expenditure, nearly $1.5 billion a day.

In almost every sphere, what America thinks and does will dominate Crunch Time. It isn't too much to say that the Crunch Time security story will be that of America's attempts to maintain its position against the other nineteen-twentieths of the world.

Keeping our distance

> [W]e have about 50% of the world's wealth but only 6.3% of its population ... In this situation, we cannot fail to be the object of envy and resentment. Our real task in the coming period is to devise a pattern of relationships which will permit us to maintain this position of disparity without positive detriment to our national security. To do so, we will have to dispense with all sentimentality and day-dreaming; and our attention will have to be concentrated everywhere on our immediate national objectives.

Only the numbers let slip that this isn't the latest memo from the White House's National Security team. In fact, it comes from PPS23, a planning document on United States policy towards Asia written in 1948 by George Kennan, one of the State Department's most brilliant officials.

Kennan went on to say:

> We should make a careful study to see what parts of the Pacific and Far Eastern world are absolutely vital to our security, and we should concentrate our policy on seeing to it that those areas remain in hands which we can control or rely on. It is my own guess ... that Japan and the Philippines will be found to be the cornerstones of such a Pacific security system and if we can contrive to retain effective control over these areas there can be no serious threat to our security from the East within our time.

America today has comparatively less wealth and less people, but the fundamental drivers of American policy remain the same: maintain the disparity. This chapter is our attempt to make sense of the increasingly complex 'security' issues that are appearing in the newspapers at the beginning of the 21st century, and the sensible response to the global picture for the individual. But our story doesn't start in the Pentagon. It starts in one of those places that Kennan thought should remain in hands that the US could control or rely on – the Philippines.

Trouble in paradise

American couple Martin and Gracia Burnham weren't used to luxury. As operatives for the zealous New Tribes

Mission, their singular purpose in life was to bring the gospel of Jesus Christ to remote people in remote places, a call Martin had been answering since childhood with his parents in the Philippines. That meant a life spent camping out in the jungle and dining on the occasional malnourished chicken.

So, when it came to celebrating eighteen years of marriage they took a break from roughing it and headed for Arreceffi Island and the five-star beach bungalows and gourmet buffets of the Dos Palmas Resort. Sitting in the tropical waters of the southern Philippines, it's a serene world unto itself. Tastefully appointed huts rest on stilts above a clear blue bay. There are picnic lunches on the water, an infinity pool and jet spa, and snorkelling on pristine coral reefs. Dos Palmas offers the complete international luxury tourist experience, removed from the everyday cares of real life. 'Discover things you never thought possible' runs the blurb on the Dos Palmas Resort website. The Burnhams certainly would.

Early on the Sunday morning, in May 2001, the Burnhams were awoken by the rough sound of armed bandits slamming a motor launch onto the beach of their resort. The gunmen headed for the tastefully appointed huts, grabbed the Burnhams, some other guests, and a number of resort staff. Corralling twenty people into the boat, they headed north, towards the scores of jungle islets that dot this part of the world.

Pursued by planes, helicopters and patrol boats, the bandits were eventually tracked to the jungle island of Basilan, where at one stage they were cornered in a hospital. They escaped, but added a local nurse to their hostage collection. Over the course of the next few months, hostages were released and killed, until the diminishing group consisted only of the kidnappers, the nurse and the Burnhams. In Basilan, it really is a jungle out there.

The capture and pursuit of the Burnhams was to become a pivotal event in an unfolding war on terror, but not because the Burnhams were in any way significant. Hostages had dominated the United States political scene since the seizure of American embassy staff in Tehran, and the kidnappings of United States citizens in the Lebanon. But the Philippines isn't part of the volatile landscape of the Middle East. International attention rapidly focused on what the most powerful state in the world would do for its captive citizens.

For a long time the answer was not a lot. Then in November 2001, six months after their capture, a local Filipino TV station aired a video clip of the Burnhams, looking exhausted and desperate. In the intervening period, nothing had changed for them. But while they were being shunted from jungle hideout to jungle hideout, everything had changed for America.

People as pawns

Around the time of the Burnhams' interview, two things were happening that would have a grave bearing on their fate. Firstly, following September 11 and America's attack on Afghanistan, the Bush administration was trying to define the notion of a 'war on terror'. They knew they had to fight terrorists other than Bin Laden's al-Qaeda, the question was, which ones? Second, the Philippines' president Gloria Macapagal-Arroyo was visiting Washington. She came with a deal. The United States could use her air and naval bases. In exchange, she wanted 'help' with a domestic terrorist problem, and about $100 million worth of military hardware.

She presented the bandit gang who'd snatched the Burnhams as a domestic terrorism problem. Actually, the kidnappers were disaffected youths from a Muslim minority who had decided to glamourise their violence with the Arabic *nom de guerre* Abu Sayyaf, meaning sword-bearer. They were led by a charismatic 40-year-old who called himself Abu Sabaya and wore Oakley wraparound shades. Originally a criminology student, Abu Sabaya had left the Philippines to work in Arabia, and had been drawn into Islamist circles that took him to Libya and Pakistan. This was enough for the Americans: the Muslim connection was clear. The Burnhams' kidnappers were no longer just local thugs, they were the new front in the war on terror and United States cash and support was quickly forthcoming.

The Philippines constitution forbids foreign forces from operating on its soil, but three months after the Burnhams' weary TV appearance, over a thousand United States 'military advisers' arrived in Basilan to train their Philippine colleagues in jungle warfare. Four months after they'd arrived the training paid off, as did the United States' other assistance: helicopter gunships and night-vision equipment. The American-backed forces cornered the gang and their three hostages in a remote ravine. The Burnhams' drama was about to end.

What to make of all this?

The question for us, just getting on with our lives, is what to make of terrible events like these, how to make sense of them in a way that adds meaning rather than confusion. The Burnhams made good newspaper copy. The story had drama and tension, and could be slotted easily into wider events in the world (the war on terror). But for Crunch

Time the issue isn't about the news qualities of the Burnhams. It's the insight their story provides into the forces that shape our lives: how we should behave in a world where you can be snatched from an idyllic holiday retreat and sucked into a local power struggle that, in turn, is just a part of a global diplomatic game.

That's where our argument began.

'This is a classic case of American imperialism', Adrian was saying over the phone one night. 'The Burnhams were an irrelevant, but fortunate, detail for the politicians.'

'The Filipinos would've got their money and the Americans their bases, Burnhams or no Burnhams', he continued. 'Remember, Bush had just bombed the daylights out of Afghanistan. He was looking for a second front, somewhere to demonstrate that the war against terror is a global fight. The Philippines was the ideal solution – a friendly country in need of some help to quell some local trouble from independence-minded rebels of an Islamic persuasion.'

'More than that, South-East Asia is somewhere strategic analysts have been twitchy about for decades. A "rescue mission" was the perfect excuse to send in troops and secure a crucial part of the American empire.'

Mike disagreed. 'You don't have to like Bush and his lot to realise that a country, especially a superpower, has a duty to protect its citizens.'

'Terrorism, even of the kind demonstrated by the Burnham's kidnappers, poses an enormous threat, not just to the United States, but to all societies that value freedom of thought and liberal ideals', he went on.

'September 11 was the worst and most shocking example, but as an activity in general, terrorism works against everything our societies stand for, it threatens lives and infrastructure, and also the way we relate to each other, our tolerance and understanding.'

'America is one of the few countries with the power worldwide to combat it. Let's be thankful for that. There's a big political game going on but policy is also about trying to protect people like the Burnhams, trying to react to September 11, the Bali bombings and everything else that happens.'

More than that, he argued, look at the facts specific to South-East Asia: two of the 9/11 hijackers had spent time in Malaysia, Ramzi Yousef, the mastermind of the first bombing of the World Trade Center in 1993, used the Philippines as a base, and it was thought that al-Qaeda might regroup in South-East Asia. In these circumstances, President Macapagal-Arroyo's invitation could be seen as a welcome opportunity for the United States to track down and destroy terrorists affiliated to the broad anti-American alliance. It would stop that alliance from gaining a foothold in a new, important and volatile part of the world and extend a helping hand to a strategic 'partner', the fledgling democratic regime that had replaced the Marcos dictatorship.

Adrian's Grand Chessboard v. Mike's 'events, dear boy'

The conversation continued over the next few days.

Adrian accused Mike of being naive, thinking that there was no grand plan. It's ridiculous to believe that President Bush and his advisers react to daily events without having some broader idea of where history is going. As if punch and counter-punch was all that matters. As if the phalanx of analysts in the Pentagon and the spin doctors who replay every move the President makes for the public have no grander plan than re-election.

He sent Mike a quote from Lord Curzon, who ruled British India at the end of the 19th century: 'I confess that countries are pieces on a chessboard upon which is being played out a great game for the domination of the world.'

He pointed to a list of American 'interventions' between 1945 and 2000: 201 different 'theatres of operation' in 55 years, according to the Federation of American Scientists. Taken together, reckoned Adrian, they look like a sustained military campaign, the exercise of imperial muscle.

Mike had to agree: taken together it does look like there's an empire thing going on here. But they can't be taken together. Anti-communism prompted the Berlin airlift. Bill Clinton's liberal interventionism took United States forces into Kosovo. Both these 'imperialist' actions were just responses to other people's moves. Stalin blockaded Berlin. Milosevic invaded Kosovo. Saddam Hussein invaded Kuwait. The United States just decided not to sit back. It did something. For which Berliners, Kosovars and Kuwaitis had reason to be grateful.

Lord Curzon was living in a different age, reckoned Mike. It's Crunch Time now, and our times are better described by 1950s Prime Minister Harold Macmillan. When asked what he feared most, Macmillan allegedly replied: 'Events, dear boy, events.' Events, not conspiracies, drive the big security issues of our time, argued Mike.

Agreeing to disagree, we labelled our respective positions.

Adrian's Grand Chessboard

The world is run by a powerful and elite group of politicians, corporate bosses, government officials and bureaucrats, assisted by the World Bank, the IMF and other odious institutions. These individuals and institutions have a single-minded objective: to maintain the disparity. They want to hang on to power and increase their own influence and wealth.

They use any means possible: military 'intervention', ownership of the media, restriction of personal rights, repressive legal systems – the lot, and internationally they enforce 'systems' that increase inequality and injustice such as Third World indebtedness, taxation systems that benefit the wealthy, and straightforward crony capitalism.

Conspiracy?

Mike's 'events, dear boy'

The world is the way it is through a combination of historical accident and human nature. People are generally well intentioned, although sometimes they make bad decisions or focus on the wrong things.

Elected politicians want their policies and reforms to benefit society as a whole, but face choices that sometimes conflict with other closely held beliefs. Our political leaders are also the guardians of our national security. One of their gravest responsibilities is to keep us safe, both now and into the future.

Business chiefs believe they are adding value to the world we live in, economically and socially. They provide jobs for employees and profits for shareholders that in turn are invested in other valuable enterprises.

Institutions such as the World Bank and the IMF mean well and their mistakes are usually visible only with hindsight. Their poor image is the result of a lazy and cynical public happy to give a credit card number but not a helping hand.

Naive?

The question we need to answer is which of these two worlds we actually live in. Which of these scenarios is more useful when it comes to making sense of global security issues on the one hand and personal crises like the Burnhams' on the other?

The strategic background

Earlier in this chapter, we met George Kennan, who wrote that the United States' main strategic focus had to be on maintaining the disparity between America and the rest of the world. That quote comes from a document known as the 'Long Telegram'. It was sent from the United States embassy in Moscow in February 1946, where Kennan was posted, to the State Department and began: 'In view of recent events, the following remarks will be of interest to the department.' Kennan's telegram proceeded to outline, in over 6,000 words, the looming threat to the Western world from the dogmatic and confrontational ideology of Soviet Russia – a clash not of civilisations, but ideologies. Stalin had outlined the confrontation much earlier – in 1927, in fact, before a delegation of American workers:

> There will emerge two centres of world significance: a socialist centre, drawing to itself the countries which tend towards socialism, and a capitalist centre, drawing to itself the countries that incline towards capitalism. Battle between these two centres for command of world economy will decide fate of capitalism and of communism in entire world.

After the Second World War, the back end of the 20th century was held together by the precarious balance of power between the United States and the Soviet Union. Staring each other down across the Iron Curtain, their mutual antagonism provided each with a focal point for their interaction with the rest of the world. The Americans fought to 'contain' communism, and communism fought to establish a grip in the regions of the world where the old colonial order was breaking down.

There were events: the Cuban Missile Crisis, the Gulf of Tonkin incident, the Soviet invasion of Afghanistan; but these were also chess moves, tactics determined by broad strategic imperatives (think *Civilization* – resources, cash, weaponry, technology, culture and diplomacy). The moves were explained in terms that referred to the fundamental ideological differences between the capitalism of the West and the communism of the Soviet Union. If Senator McCarthy was hunting down communists in Washington DC, it was only logical that the President should be hunting them (or at least 'containing' them) abroad.

Armageddon was avoided. Perhaps precisely because of the mutually assured destruction of a nuclear conflict between two powers with a combined arsenal powerful enough to destroy the entire planet 50 times over. Or perhaps not, maybe we were just lucky. In any case the indisputable fact, after the fall of the Berlin Wall in November 1989, was that America emerged victorious from this decades-long confrontation. For many commentators, the US won not because of the brilliance of its strategic planners, and not because of the superiority of its military prowess, but because of the inherent ability of its capitalist economic system to produce growth, against the sheer political strain of maintaining a totalitarian state. The Soviet Union simply couldn't afford to spend as much money on arms as the United States. It had enough trouble getting its bakers to bake bread.

Cause for celebration for the West as this may have been, when the Soviet Union collapsed, so too did the stabilising, if repressive, influence of the Cold War and its reassuring Manichaean take on the world.

From Russia in the West, through to China in the East, states in central Asia that had fallen squarely under the influence of the communists were now free to bargain with whomsoever they liked. And they had a lot to bargain with.

It's against this background that the events of the early 21st century have played themselves out so far, and it's this background that we must use to analyse which of our theories is closer to the truth.

A game of chess

History is the sum of big deeds and small. Sometimes it takes distance to determine which remain truly historic, decisive captures on the chessboard, and which were just noise.

When Austria's Archduke Franz Ferdinand and his wife were shot dead by the Miljacka River in the centre of Sarajevo, it began the chain of events that led to the First World War. That was historic. Later in the same century that river flanked the no-man's-land off Sarajevo's 'sniper's alley'. Many more people were shot dead, but no global war ensued. That was noise – tragic and heartbreaking, but just noise.

The difficulty as we read the news each day is determining what is historic, and what is just noise.

When Archduke Ferdinand was assassinated on a

bridge in Sarajevo, kicking off World War I, it was – from this distance – a defining moment in the 20th century. But even from much closer, whichever way you cut it, September 11 of 2001 was an equally momentous day: a day that altered the entire strategic background of Crunch Time. When hijacked aircraft destroyed Manhattan's tallest buildings and took a chunk out of the Pentagon, they destroyed very publicly the myth of the American heartland's invulnerability.

On 10 September the Bush administration was facing flak for pressing for oil drilling in Alaska, and a big part of the President's tomorrow was to be spent reading to a kindergarten class. 9/11 was a classic 'event', of a kind that even Macmillan's worst nightmares couldn't conjure up. Before September 11, the Bush administration had been happily pursuing business relations with the Taliban and Saddam Hussein was just an irritating hangover from the first Gulf War. Suddenly, the world changed. The Taliban were harbouring an international terrorist mastermind, and Saddam Hussein was developing weapons of mass destruction, an imminent threat to the West. Checkmate, Adrian. Game, Mike?

Unfortunately, nothing's quite so simple.

When the Bush administration decided that 9/11 was the work of al-Qaeda, they decided to go after its head, a renegade Arab millionaire called Osama bin Laden. According to their best intelligence, bin Laden was based in Afghanistan where he had once operated alongside CIA-backed proxies advancing an earlier strategic goal of America's – fighting the Soviet Union.

Afghanistan sits between one of the biggest new supplies of oil in the world, the Caspian, and two of the biggest new markets, India and Pakistan. 9/11, like the kidnapping of the Burnhams, might've been an event out of the blue, but the players of the Grand Chessboard could certainly use it to their advantage.

Bill Clinton says he had the Pentagon draw up plans to invade Afghanistan years before, after al-Qaeda had attacked the USS *Cole* in 2000, but he couldn't use them because the US didn't have basing rights in Uzbekistan. After 9/11, though, when Bush and his team went knocking on Uzbekistan's door, the country opened its arms to the American infidel. The 'events' were a gift to an administration which knew only too well the significance of that part of the world.

Speaking not as the Vice-President of the United States but as a consultant on oil pipelines to central Asian countries, Dick Cheney said at a conference in 1998: 'I cannot think of a time when we have had a region emerge as suddenly to become as strategically significant as the Caspian.' That word 'strategically' is Washington code for 'militarily'.

Zbigniew Brzezinski, strategic adviser to several United States presidents and one of Washington's pre-eminent geopolitical gurus, reckons Eurasia has been the centre of world power since the continents started interacting some 500 years ago. His thinking, consciously echoing Lord Curzon's, is summed up in *The Grand Chessboard: American Primacy and its Geostrategic Imperatives*. It's *the* book if you play world politics like chess, and, according to Adrian's Grand Chessboard theory, that's just how they play it in the powerful corners of Washington.

Control of resources

There may be no checkmate in global chess but there are rules. Control of resources is rule number one, and resources are only resources to the resourceful. Access to large supplies of timber for mast construction was vital for the British Navy in the age of sail. Coal supplies were crucial in the age of steam. When the British fleet switched from steam turbines to diesel in the run-up to the First World War the strategic necessity was access to oil.

Oil now underpins every facet of our lifestyle. Nothing wrong with that. So too does China's future growth. And India's. And Brazil's.

The energy that keeps our shipping lanes, airports and roads busy today doesn't come from burning wood or coal, but from oil. Energy security requires control of the geographic area where the largest supplies of oil currently reside (the Middle East) and of Eurasia, where the world's remaining supplies are waiting to come online.

This view of the world says that dominance of central Asia not only will ensure new sources of energy and mineral wealth, but will also form a watchtower over the oil reserves of the Middle East. Remarkably, Brzezinski's book reads like *Civilization*'s 'User's Guide': 'The three grand imperatives of imperial geostrategy are to prevent collusion and maintain security dependence among the vassals, to keep tributaries pliant and protected, and to keep the barbarians from coming together.' The Romans couldn't have put it better.

Right in the middle of Eurasia sits the Caspian Sea, packed full of oil. Pipelines from the Caspian must travel through neighbouring countries – specifically through Russia (now a 'friend' of the United States, but not a trusted one), Iran (a nuclear rebel and member of George W. Bush's 'axis of evil'), or Afghanistan (to get to Pakistan or India).

Forget Events, we're back to the Grand Chessboard.

Despite America's immediate invasion of Afghanistan, the Afghans weren't the most obvious culprits for the 9/11 crimes. None of the 9/11 hijackers were Afghans (or Iraqis, for that matter). Fifteen of the nineteen hijackers were dissidents from a Gulf state named after the family that ran it – Saudi Arabia.

Saudi Arabia is the spiritual home of one of the world's great religions, Islam, is immensely rich, has the world's biggest oil reserves, and at the time was also a major Middle Eastern base for American soldiers and airmen (one of the factors that so enraged the hijackers). The events of 9/11 put the superpower into a compromising position: the connections between Saudi Arabia and the terrorist events were clear, but the United States could hardly point the finger of blame at the internal politics of the world's biggest oil producer.

Indeed, when the United States House-Senate Intelligence Committee released its 900-page report on the 9/11 attacks in July 2003, the 28-page section relating to Saudi Arabia was removed. It had been classified.

However, the Bush administration decided it didn't have to get heavy with the Saudis at that stage: for the grandmasters of world chess the fact that Saudi renegade Osama bin Laden was encamped in Afghanistan after the 9/11 attacks was a gift. The fundamentalist Muslim Taliban government in Afghanistan had already annoyed US policymakers because it wouldn't do as it was told, and its repressive policies to women had alienated US politicians. It couldn't be trusted to allow a valuable oil pipeline through its territory, and it occupied the eastern flank of a regional threat – Iran.

So, a war instigated by dissidents from a strategic ally – Saudi Arabia – meant the removal of United States troops from that country, jeopardising American power in the region. But 9/11 then took them to Afghanistan, then on to the Middle East's major regional irritant, Iraq. Control of Iraq, regardless of the problems, gave the United States the direct access to Saudi Arabia that it had forfeited. It also meant that the last remaining independent regional power, Iran, faced potential threats from Baghdad to the west and Kabul to the east. The war on terror suddenly had a logic. Strike out. Secure resources. Establish bases. The Philippines was just a bit player in the grand scheme designed to maintain the disparity. Forget 'events, dear boy'. Security is a chess player's game.

Non-events

Revealingly, it's more difficult to find evidence to support the events theory – but it can be done. The key to understanding lies with the concept of democracy. If America and its allies stand for anything, says this point of view, it's freedom: the right to pursue your own dream, whatever that dream might be, and the energy, vitality and growth that freedom brings to the world as a whole. Democracy protects people's right not to be told what to do by the state: if we don't like what they're up to, we can kick out the chess-playing elite.

So America must pursue the cause of democracy, not only in the protection of its own people's rights to democracy, but also, pre-emptively, wherever freedom is threatened by outside interests. By attacking states that provide succour to terror, America is protecting the free world from its enemies. It's making the world secure for us to pursue our individual dreams.

By confronting al-Qaeda in Afghanistan, Saddam

Hussein's Iraq, the gangsters of the southern Philippines, and – as the 21st century progresses – whoever else looks dangerous to American interests, America is rightfully defending the security and freedom of everyone who shares its values of freedom and justice.

Harvard professor Samuel Huntington – the author of *Clash of Civilizations* and whose name (along with Francis Fukuyama's) invariably surfaces whenever big geopolitical issues are discussed – comes down on Mike's side of the fence.

> A world without United States primacy will be a world with more violence and disorder and less democracy and economic growth than a world where the United States continues to have more influence than any other country in shaping global affairs. The sustained international primacy of the United States is central to the welfare and security of Americans and to the future of freedom, democracy, open economies and international order in the world.

Of course, the challenge for Huntington's view is communicating the good the US is bringing the world over the sound of the TV news announcing the accidental bombing of a Lebanese wedding party, or the progress of an Iraqi child amputee. But in contrast, devotees of the chess theory point out that, since the media is owned by members of the power elite, it's all too easy to drown out the tragic consequences of their heinous actions under Orwell's diet of sport, crime and astrology, not to mention sex and Big Brother-style reality TV.

Still, to a greater or lesser extent, the images of destroyed lives do make it through the noise, and we inevitably ask whether our ends justify our brutal means. It's certainly enough to make you question the motivations of our leaders.

Are you confused and overwhelmed?

> *Philosophers must become kings in our cities or those who are now kings and potentates must learn to seek wisdom like true philosophers, and so political power and intellectual wisdom will be joined in one. Until that day, there can be no rest from the troubles for the cities, and I think for the whole human race.*
>
> Plato

You're not alone.

The complexities of governing a sprawling and threatened Empire have been common to governors since the dawn of time. Plato's solution was to create 'philosopher-kings': a ruling class in a just society should be men apprenticed to the art of ruling, drawn from the rational and wise, bred from the finest families. Government, he said, was a special art in which competence, as in any other profession, could be acquired only by study. Strangely enough, we have more students of government these days, but few, if any, philosopher-kings.

But we needn't feel too bad if our leaders don't match up to Platonic ideals. Some of the greatest minds of the 20th century grappled with this issue in vain. In 1932 physicist Albert Einstein wrote to psychologist Sigmund Freud, lamenting the inability of the world's brightest and best to influence the course of world affairs and asking him how intellectuals could work towards peace.

Einstein revealed himself a confirmed adherent of the

Grand Chessboard theory. Those who crave power are hostile to any notion that would limit national sovereignty, said Einstein, and a 'small but determined group' who benefit financially from war supports them in this. They regard warfare 'simply as an occasion to advance their personal interests and enlarge their personal authority'. How, Einstein enquired of Freud, is it that men are aroused so easily into hatred, 'even to sacrifice their lives'? And, more ambitiously, 'is it possible to control man's mental evolution so as to make him proof against the psychosis of hate and destructiveness?'

Freud's response? Don't hold your breath.

'Conflicts of interest between man and man are resolved, in principle, by recourse to violence', he wrote. Even the law is a form of violence – 'the suppression of brute force by the transfer of power to a larger combination, founded on the community of sentiments linking up its members'. Men are brutal victims of their urges, said Freud, urges of love and urges of hate, of conservation and unification versus destruction and murder. His solution?

> A superior class of independent thinkers, unamenable to intimidation and fervent in the quest for truth, whose function it would be to guide the masses dependent on their lead. There is no need to point out how little the rule of politicians and the Church's ban on liberty of thought encourage such a new creation. The ideal conditions would obviously be found in a community where every man subordinated his instinctive life to the dictates of reason.

Former academics like Zbigniew Brzezinski and Henry Kissinger perhaps? Well, the world isn't run from a university campus yet, and you don't hear a lot of public debate suggesting that it should be.

In fact Brzezinski, the old chess master, was so devoted to the idea of a bipolar world, where America's every move as white countered each Soviet move as black, that he, like countless others, was blind to the forces they had helped call into being. In Afghanistan, the Islamic 'patriots' who fought the Red Army would turn their venom on America and its allies.

As for 'reason', one solid lesson Freud's own century has left us with is the inadequacy of reason as the sole compass for our affairs. We have seen that using reason humanity can justify anything, from genocide to global warming. Reason needs to accommodate our humanity, with its fear, compassion and hatred.

Can Mike and Adrian agree?

First and foremost, the stark difference between Mike and Adrian's respective interpretations emphasises the fact that we have a duty to scepticism. When the crunch comes it's our responsibility to question, examine, interrogate and judge the motivations of those in power. More than that, we should have the ability, when we feel necessary, to act to influence them.

If 9/11 opened many people's eyes to a world they'd chosen to ignore, it's also opened them to the fragility of our comfortable existences. In the 21st century, scepticism must become a core competency. The events of September 11 were a bloody and extraordinary *coup de théâtre* intended both for an audience of 1.2 billion Muslims and as a ghastly modern nightmare for the rest of us. That single act resulted in the deaths of nearly 3,000 people.

Over the next year, the United States Congress gave the Pentagon $28.5 billion to spend on emergency measures. By contrast, in 2001 more than 30,000 Americans committed suicide. It remains the eleventh leading cause of death in the United States. Over the next year, the United States Department of Health and Human Services gave just $2.5 million to establish a national suicide prevention resource centre. In case the millions and billions are confusing, that's around a thousandth of the money that went to the military – excluding the campaign in Afghanistan. In 2002, terrorists killed 30 United States citizens. The same year more than 31,000 Americans committed suicide.

A little over a year after their capture, and a few days after their 19th wedding anniversary, the Burnhams were caught up in a desperate firefight. Martin Burnham was shot and killed. His wife Gracia was the only survivor, and she was badly wounded. But the Filipino military was happy: their troops had killed Abu Sabaya, although with no dead body to be found they had to make do with waving his captured Oakleys about.

That was the tragic end to the Burnhams' weekend in paradise. The international luxury tourist experience had lost in a face-off with real life. For these innocents, events lost out to the Grand Chessboard. And Abu Sayyaf hasn't gone away.

For us, sitting outside the game, 'security' relies on assumptions which, when looked at closely, often turn out to be illusions. Military might is a necessary component of power, and one that still elevates the state above commerce. But the opportunities to wield it to defend the resources that underpin our way of life are neither simple nor necessarily effective. Democracies don't always want to be defended. And when we drive our gas guzzlers to anti-war demonstrations we reveal, only too clearly, our ability to juggle contradictions that policymakers, in the glare of public scrutiny, find less simple to resolve.

CHAPTER 8

Barbarians at the Gate

Childless countries and the fear of an immigrant future

In the second half of the 4th century a nomadic people from what's probably now Korea began a long journey west. Whatever their reasons for leaving their eastern steppes and pastures, they kept going decade after decade until they reached the edges of Europe. They fought with bows and arrows and travelled on horseback. They were called the Huns.

History is silent on their progress until they reached the borders of the world known to the Western Roman Empire. By the third quarter of that century they'd reached the plains between the Dniester and Dnieper rivers, in what's now the Ukraine. The people who lived there were called the Greuthingi. The Huns crushed and enslaved them.

Not long after, they turned their attention to the Greuthingi's neighbours, the Wesi, sacking their settlements, raping and pillaging their way along the great river boundary of the Western Roman Empire – the Danube. The Wesi's leader, Fritigern, sent a diplomatic mission to the Roman Emperor Valens, asking for help. Fritigern persuaded a bishop, who had previously built his reputation saving Christians from the Wesi, to head the mission. Faced with the superior military power of the Huns, the Wesi wanted asylum status for their entire people under the protection of Rome.

Reluctantly Valens granted permission for the refugees to cross the Danube and settle in the Empire en masse. Tens of thousands of people and their possessions came west as immigrants.

By all accounts, benefits and social services for the newcomers were a little thin on the ground. The Romans, according to the great historian Edward Gibbon, 'levied an ungenerous and oppressive tax on the wants of the hungry barbarians. The vilest food was sold at an extravagant price and … the markets were filled with the flesh of dogs and unclean animals who had died of disease.'

Fritigern's people turned on the Empire that had taken them in. At Adrianople, on what is today Turkey's European border, the Wesi routed and butchered Valens and his Roman legions. Fritigern's successor took the refugees on to Rome where they sacked the city and started the process we know as the decline and fall of the Roman Empire.

Fritigern's people are better known as the Goths, the original barbarians at the gate. Today their modern equivalents camp outside the Channel Tunnel. The Goths arrived in the Empire as immigrants. They were the enemy camped beyond the walls, waiting for an opportunity to turn on their hosts and masters and bring down the glory of civilisation. Yikes.

Our Crunch Time hordes

What becomes of the surplus of human life? It is either, first destroyed by infanticide, as among the Chinese and Lacedemonians; or second it is stifled or starved, as among other nations whose population is commensurate to its food; or third it is consumed by wars and endemic diseases; or fourth it overflows, by emigration, to places where a surplus of food is attainable.

James Madison, 1791

Let's be clear about this: we live in a divided world. On one side of the line sit the 'golden billion', those lucky enough to be born in the 'rich' world. On the other side of the line sit the rest. Five billion of them.

The line that separates these two global constituencies is dotted. Millions of people try and cross it every year. Some succeed, some die, some get sent to hellish housing estates in Glasgow by the British government, some get sent to prison on hellish Pacific islands by the Australian government, others aren't so lucky – they get sent home.

Those of us born into the golden billion stand a good chance of living a very long life in relative comfort, meeting and exceeding our need for calories, warmth and entertainment. Those born on the wrong side of the golden line stand a good chance of not meeting any of those requirements.

Globalisation is making a 'winner take all' world. In 1870, the world's richest countries, Britain and the US, had incomes per head roughly nine times larger than the poorest countries. In 2002, that number had shot up to 225. At the beginning of the 21st century, the richest 15 per cent of the world has almost 80 per cent of the world's wealth, while the poorest half of the world gets by on just over 1 per cent.

This gap, between those who could clearly afford to worry less about their standard of living, and those who don't know what standard of living means, only got bigger over the last century. The World Bank calculates that the average income in the world's richest 20 countries is 37 times the average in the poorest 20, a gap that doubled in the last four decades of the 20th century.

The UN reckons that nearly half the world's population lives on less than US$2 per day. Asylum seekers may not know exactly what lies in store for them on the other side of that dotted line, but their economic decision-making skills are as rational as you can get. On the rich side of the line, even the cows earn more: in Europe, each cow receives $2.20 a day from the taxpayer in subsidies and other handouts, while in Japan, cows make an even better living, getting $7.50 a day.

But these are just numbers. To say, for example, that a billion people live on less than a dollar a day is one thing, to understand what that life is like is another. The World Bank says it means 'malnutrition, illiteracy, disease, squalid surroundings, high infant mortality and low life expectancy'. In other words, nasty, brutish and short.

More ugly statistics. Of the world's billion or so absolute poor, the World Bank says two-thirds are underfed, and the same number illiterate, and almost all are drinking dirty water with no access to even the most basic toilet facilities. In rich countries, less than one child in a hundred dies before the age of five, in poor countries a fifth of all children will die before that birthday is reached. Average life expectancy in rich countries is pushing 80; in sub-Saharan Africa, it's around 50.

Meanwhile, *Forbes* magazine announced in September 2006 that the net worth of the 400 richest people in the US was $1.25 trillion. The two richest, Bill Gates and Warren Buffett, have assets that exceed the combined gross national products of 600 million people in 49 of the world's poorest countries.

When the people held in detention camps or living illegally consider whether they ought to leave their homes, lives and families and undertake life-threatening and treacherous voyages in an attempt to cross that line, they're acting on the basic human impulse to improve the hand they were dealt at birth. Experts at the Rand Corporation think tank in the US reckon that migrants who score a green card letting them into the country increase their lifetime earnings by about $300,000. It's a lottery win.

It's no mystery why politicians and the supporters who vote them into office behave as if they are engaged in a war with the barbarian hordes. They are.

In this age of globalisation, information may flow freely around the world; capital sloshes in bucket loads around the globe, goods and services too. But people don't. Politicians play to their constituents' fear of 'being swamped' (to use the words of Britain's former Home Secretary) by the unwashed hordes on the other side of the dotted line. And the situation isn't going to get any better.

The winner's curse

For most species, evolutionary success is measured in sheer numbers. If a particular type of fly or baboon is successful against predators, if it's good at finding food and water, surviving weather and natural disasters and at defeating disease, its numbers will grow.

By this strictly biological definition, *Homo sapiens* has been immensely successful, particularly over the 20th century. Despite two world wars, genocides, persecution and conflict, the number of people on the face of the earth grew from around 1.5 billion people in 1900 to more than 6 billion in 2000, a fourfold increase. When we passed the 6 billion mark (around 12 October 1999 say demographers), *Homo sapiens* had already exceeded the biomass – the sheer bulk – of any large animal species that ever existed on the land.

We win.

There is, however, a noisy bunch of people who think this is a bit simplistic. They reckon the human race is too successful for its own good. Their argument goes like this ...

The world's 6 billion people are already chewing up the earth faster than they can repay the debt. The scientists who measure these things think we passed the point where we were using all the resources that nature could replenish back in 1978. Ever since then, we've been draining the tank, exhausting fisheries, fossil fuels, water supplies. The lot.

Some scientists reckon that, if we're all to have European standards of living and enough food and water to go around, the earth can sustain only about 2 billion people. So, the question becomes, who's jumping off, you or me? For the rich world politicians and their electors – that's us – the answer is clear: them.

Looking at this all-too-easy-to-paint picture of doom and gloom, there's at least one figure from history who would sport a wry smile.

Too many people, not enough food

In Europe at the end of the 19th century, the civilised world of dinner parties, gentlemen's clubs and genteel philosophy was agog with the possibilities afforded by the French Revolution. Here, it seemed, was a blazing example of a modern democratic politics in action. Utopia here we come. French-style democracy would take the world to the next level of civilisation in which people conducted politics in a civilised way without resorting to violence, a social order that could provide everyone with opportunity and sustenance.

In France, for instance, the Marquis de Condorcet, a liberal aristocrat, wrote an essay arguing that human progress unfolded naturally in a series of ten stages, the ninth ending with the founding of the French Republic, the tenth and last being a world of equality in wealth, gender and opportunity, and, crucially, a world of abundance.

Condorcet was not alone in his optimism. Others picked up the revolutionary theme and ran with it. In England, for instance, writer William Godwin (poet Shelley's father-in-law) wrote about the forthcoming society of equality, peace, happiness and altruism in a bestseller called *Of Avarice and Profusion*.

For progressives like Condorcet and Godwin, a rosy new era was dawning. Conservatives, on the other hand, were frightened by the signs of social chaos and decay they saw all around them and the spectre of social collapse. According to one of the most influential of their number:

> The great and unlooked for discoveries that have taken place of late years in natural philosophy, the increasing diffusion of general knowledge from the extension of the art of printing, the ardent and unshackled spirit of

> inquiry that prevails throughout the lettered and even unlettered world, the new and extraordinary lights that have been thrown on political subjects which dazzle and astonish the understanding, and particularly that tremendous phenomenon in the political horizon, the French Revolution, which, like a blazing comet, seems destined either to inspire with fresh life and vigour, or to scorch up and destroy the shrinking inhabitants of the earth, have all concurred to lead many able men into the opinion that we were touching on a period big with the most important changes, changes that would in some measure be decisive of the future fate of mankind.

Sound familiar? This Crunch Time philosopher was Thomas Malthus, an English cleric who sat down, he said, to write up some thoughts following a conversation with a friend about 'the general question of the future improvement of society', and ended up kicking off a centuries-long feud about the nature of human society. Malthus was a pillar of England's conservative establishment and the ideas of Condorcet and Godwin about the future of society threatened to undermine everything he stood for. His answer was the none-too-snappily-titled *Essay on the Principle of Population, as it affects the Future Improvement of Society with Remarks on the Speculations of Mr. Godwin, M. Condorcet, and other Writers*.

Malthus opened with a couple of ideas that for him were a given.

> First, that food is necessary to the existence of man.
>
> Secondly, that the passion between the sexes is necessary and will remain nearly in its present state.

Reasonable enough. He then went on to draw a conclusion which has shaped the nature of the population debate ever since.

> The power of population is indefinitely greater than the power in the earth to produce subsistence for man.
>
> Population, when unchecked, increases in a geometrical [1,2,4,8…] ratio. Subsistence increases only in an arithmetical [1,2,3,4…] ratio. A slight acquaintance with numbers will shew the immensity of the first power in comparison of the second. This implies a strong and constantly operating check on population from the difficulty of subsistence.

As a result, wrote Malthus, French Revolution or not, humankind was destined to be caught in a struggle between a human population continually trying to expand, and a world unable to feed it. Additional checks on human population would be provided by war, disease, poverty, famine and crime. Malthus published his essay in 1798 and parachuted himself into instant celebrity with these doom-laden assertions.

Ever since he's been reviled or praised, but never ignored by all sides in the population debate, and had his thinking bent to all sorts of purposes.

Doomsayers reckon Malthus' day will soon be with us. In the 1960s and 1970s, a rash of ultra-Malthusian books announced that explosive population growth would soon overwhelm the earth's ability to feed its people. Disaster was imminent. In 1968, biologist Paul Ehrlich published *The Population Bomb*. If the title gave a pretty good idea of which way Ehrlich thought things were headed, his opening made it crystal clear. 'The battle to feed all humanity is over. In the 1970s and 80s hundreds of millions of people will starve to death in spite of any crash programmes embarked upon now.'

At the same time, ecologist Garrett Hardin, who first coined the term 'tragedy of the commons', applied it to population growth and its effects on the earth's finite resources. He argued that you needed to combine political and social theory with biological data. 'A finite world can support only a finite population: therefore, population growth must eventually equal zero.'

Like Ehrlich, he'd seen the future and come to tell the rest of us that it sucked. Hardin wanted rigorous regulation of the human population: 'Freedom to breed will bring ruin to all.'

In the 1990s Paul Hawken continued the theme in *The Ecology of Commerce*. 'People are breeding exponentially. The process of fulfilling their wants and needs is stripping the earth of its biotic capacity to produce life; a climactic bust of consumption by a single species is overwhelming the skies, earth, water and fauna.'

Other books and articles followed the same line, with titles like *Born to Starve*, and *The Ostrich Factor*, academics and commentators warned that feeding the masses was going to be impossible. Each of these invoked the ghost of Malthus, saying that after centuries of growth, the impending disaster represented by the population boom of the late 20th century was the old vicar's warnings come to life.

They couldn't have been more wrong. According to the United Nations Food and Agriculture Organization, the number of hungry people in the developing world dropped by more than 120 million between 1980 and 2000, even though the world's population grew by 1.5 billion over the same period. A recent study reckons somewhere in the region of 25 million people starved to death in the two decades since Ehrlich wrote his book. Nothing for the world to be proud of, but far from the hundreds of millions that Ehrlich predicted.

This gives powerful fuel to the population optimists, who point at the doomsters' predictions and laugh. Look at the vast growth in human population over the 200 years since Malthus was writing, they say. We have managed all right so far, no reason to think we might not manage so well in the future.

The human race, point out the pop-optimists, has been quite successful producing enough food for its growing needs over the 200 years since Malthus was around. And most of this success has been due to our newfound ability to grow more food from the same amount of land, through mechanisation, fertiliser, weedkiller, crop rotation, new seed varieties, and all of the other great things we have learned in the agricultural revolution of the last two centuries.

But the pendulum is swinging Malthus' way again, and it's Africa where the reverend's logic is being revived. Between 1995 and 2000, cereal production in sub-Saharan Africa went up by 4 million tons to 73 million. The number of people living there went up by 76 million to 653 million. Back in 1995 that worked out at 330 grams of cereal per person per day. In just five years that was down to 300 grams. Thomas Malthus winks down through history: 'Don't say I didn't tell you so.'

While African children go hungry, in the rich world (beyond the odd carbohydrate craving) most people have forgotten what it's like to be really hungry. In fact, the rich world is suffering badly from the opposite problem – we're too fat. Lazy lifestyles and overeating combine as one of the biggest causes of serious health problems. According to the World Health Organization, they lead to the deaths of 2 million people a year across the world. Obesity is the main cause of heart disease – the world's biggest killer (more than AIDS, malaria or war); is the principle risk factor in diabetes; and is heavily implicated in cancer and other problems. In just the five years between 1995 and 2000, the number of clinically obese people in the world ballooned to 300 million, over a third of who are thought to be suffering from weight-related health problems. We did say that we live in a divided world, didn't we?

The numbers game

The global population has doubled in the last 40 years.

When Adrian was born there were 3.2 billion people on earth. It has increased by 97 per cent since then. When Mike was born there were 3.4 billion people on earth. And it has increased by 82 per cent since then. Here are the countries with the highest average number of children per woman, in 2006.

1. Niger 7.46
2. Mali 7.42
3. Somalia 6.76
4. Uganda 6.71
5. Afghanistan 6.69

Every second, five people are born and two people die, a net gain of three human beings. At this rate the world's population should double every 40 years. So, by the end of the century there would be 36 billion of us, which would be a squeeze. But the rate is slowing down. The UN reckons that the world's current population of over 6 billion will, by 2050, be between 7 and 14 billion – with a best guess of around 9 billion. Then it will start to fall.

The statistics buffs who model these things call this process 'demographic transition'. It starts with people living longer lives, as society provides better health care, hygiene and nutrition – death rates go down. For a while, birth rates stay high while death rates fall, causing a baby boom. But eventually social pressures shrink the birth rate back down and the population stabilises again. Economists point to the ability of parents to invest more in fewer children leading to smaller families. This pretty much describes what's happened in Europe during the three centuries of the industrial revolution.

That whole process has been accelerated in developing countries, especially East Asia, where it's been telescoped into decades rather than centuries. Today all the nations of East Asia, with the exception of Mongolia, have below replacement rate fertility. After galloping expansion (700 million more Chinese have joined the world since the 1950s) China now has two territories, Macau and Hong Kong, with the lowest birth rates in world.

Today about half the world lives in countries with sub-replacement fertility. The European Union is below replacement. We might fear being overrun by Eastern European neighbours, but they too are all dramatically below replacement fertility. The same is true for Turkey and Ukraine. And in the Middle East, Iran, the Lebanon, Tunisia and Algeria all have sub-replacement rates.

Population changes take a lifetime to filter through. People born today will have children some 20 to 40 years from now, and with new technology possibly even later. As women have children later in life, population growth slows. Women born today will still be fertile and having kids until the 2040s. Even so, it takes a long time for demographic changes to really have an impact. Italy's birth rate is just 1.28 per woman, but the actual population will barely fall by a few million until well into the century. In the meantime, the population balance will shift; the country will get more elderly people, and fewer *bambini*.

Those shrinking population blues

Catholic Italy, as we mentioned above, has an extremely low birth rate, one of the lowest in the world. It's so low that in 2002, in an historic address to the country's parliament, Pope John Paul II begged Italians to have more kids. He knew the birth rate in the Vatican City wasn't going to swing it.

Politicians are also taking notice. In 2006 Vladimir Putin, President of Russia, chose his annual national address to identify the single biggest problem facing his country: shrinking population. Russia's population is declining by 700,000 a year because of low birth rates, high mortality and migration. If Russians vote with their fertility, in a couple of hundred years time there won't be any Russians left.

Singapore's Prime Minister Lee Hsien Loong chose a National Day Rally to upbraid his fellow citizens on the dismal fertility rate of their tiny city state. In the 1970s Singapore ran a campaign called 'Two is enough!' aiming to put the brakes on soaring population. But at current rates, there'll be just 1.5 million Singaporeans by the middle of the century, half the current number. The government's growth targets mean they'd need nearly 4 million immigrants to keep the population on track. Singaporeans are facing up to being swamped, or just disappearing from view.

The Spanish have been fretting about this for a while. Their fertility rate has been the lowest in the world and currently bumps along near the bottom of the league table. Spanish politicians of all persuasions are worried. Generally middle-aged and heading towards retirement, they can see a problem coming: who is going to pay their pensions? Low birth rates mean a shrinking labour force and a smaller tax base from which to fund social services, including pensions. Demographers reckon that by 2050 Spain will have the world's highest percentage of the elderly – 44 per cent of Spaniards will be over 60.

So why aren't Spanish women having kids? Academics say it's all down to women's greater contraceptive and life choices, meaning more women choose to delay childbirth, limit family size or remain childless. But this isn't what Spain's women say. According to a survey by the Spanish Institute of Statistics, nearly a third of Spanish women of childbearing age want more *niños*. Asked why they don't have them, Spanish women have different reasons to the demographers: high cost of living, high female unemployment, low pay and job insecurity, employer hostility to maternity and lack of affordable, accessible and high-quality childcare. Studies in Germany and Britain suggest the same. Women would have more kids if they could. That's the 'rich' world for you.

But Spanish politicians aren't struggling to address these issues. Their solution is more blunt – they're offering a cash bonus to women who have a third child. In 2004, Australia introduced baby bonuses, saying it wanted Aussies to have one for mum, one for dad, and one for the country. Australian parents get about £2,000 per child, and early signs are that it's prompted a modest up-tick in the fertility rate. Nevertheless, state bonuses for babies haven't worked for Singapore or Japan.

There is another solution for Spain, a country whose southern tip lies just twelve miles from Africa, but it's the spectre of this which terrifies the country's leaders: immigration. Spain has some of the strictest immigration laws in Europe: immigrants make up just over 3 per cent of the Spanish population. Letting people in could fund the pensions and the social services. But Spaniards don't want to grow old in a country full of foreigners. They'll do anything other than let in the Goths.

Double jeopardy

So there are two looming issues. Both are with us now but the span of human life means they will impact far into the century. There will be a few billion more people born into the poorer parts of the world. But if literacy improves and incomes increase, then more women around the world will likely take the same decisions of those in developed countries and refuse to accept the risks and demands of motherhood. The future will be a declining human population.

If you don't allow immigration, the size of the working-age population shrinks faster than the overall population. Because of this you need more immigrants to stave off a drop in the working-age population. When the UN looked at this issue in a report called 'Replacing Migration', it came to a stark conclusion. To stop developed societies ageing at all – to keep them at the levels they are at today – would require such massive immigration that by 2050 populations of countries that chose that route would consist of between 59 per cent and 99 per cent of new immigrants and their descendants. 'Swamping' might preserve the age structure of our societies but it would transform them beyond recognition.

Already it's inevitable that populations across the developed world will age rapidly in this century. In modern welfare states, the cost of caring for the elderly falls to the government and family. An increasing proportion of older people means an increasingly expensive state pension and health service, and the only room for that budget to grow is through higher taxes on a shrinking workforce or a dramatic cut in benefits or retirement age. The further the fertility rate falls, the greater the incentive for people to have fewer children.

The advanced societies we have both inherited and created with their consumer values and institutional designs are places where people have chosen not to reproduce. And that's quite literally killing our future.

CHAPTER 9

Democratic Dreaming

The promises and parameters of politics

> *The individual man does not have opinions on all public affairs ... He does not know what is happening, why it is happening, what ought to happen. I cannot imagine how he could know, and there is not the least reason for thinking, as mystical democrats have thought, that the compounding of individual ignorances in masses of people can produce a continuous directing force in public affairs.*
>
> Walter Lippmann

American journalist Walter Lippmann wrote that back in 1925. A powerful whiff of cynicism still curls from each sentence like smoke from a Gran Corona. Born at the end of the 1880s, Lippmann started life a socialist. But by the 1920s he'd concluded that the modern world was just too complex and people just too dumb – not only for socialism, but also for democracy.

It wasn't just getting older and grumpier that changed his mind. Lippmann's cynicism was triggered by doing patriotic work for President Woodrow Wilson, getting his fellow Americans excited about joining the First World War. Serving on the presidential propaganda committee impressed on Lippmann that democracy was not about rational debate producing reasonable policies. It was all about policymakers' realities and the public's perception of them.

He had realised what democrats in ancient Greece knew only too well – that people's better judgement could be swayed more easily by emotion than by reason. The Greeks called it demagoguery, and it could, on occasion, get you kicked out of town. The Victorians called it rabble-rousing. One of Lippmann's fellow propagandists, Edward Bernays, coined a modern, less awkward term for it: public relations.

Success in kicking the Kaiser and selling the trenches to the American public didn't give Lippmann a warm fuzzy feeling. Like a comedian contemptuous of audiences willing to laugh at cheap material, Lippmann came to despise the 'bewildered herd' he'd helped mobilise for war.

We're all part of Lippmann's bewildered herd now. For those of us with aspirations to participate in democracy (or at least to feel as if we're participating), Crunch Time has compounded old problems and added new ones to those that Lippmann identified a century ago, problems that go to the very heart of our democratic culture and threaten to make a farce of our claims to a civilised political culture. But the issue common to Lippmann a century ago and to us today is that we can't disengage – democracy, for better or worse, is all we've got.

Why does democracy matter?

Democracy doesn't matter, but government does. To government belongs the exclusive right to force, and what it can (or can't do) impacts on all of our Crunch Time issues in the most fundamental ways. No matter how much energy any one of us puts into moulding our world, a government edict can destroy those efforts with the stroke of a pen or the swing of a truncheon. To get to grips with the challenges that our societies will be facing in the 21st century we need to look at the system behind the pen stroke and the riot shield.

Arguing against democracy is like picking your nose in public: nice people don't do it. Democracy is our bottom line, if only because, as Churchill so famously observed, it's the worst system, except for all the rest.

But democracies don't have a particularly impressive track record. The rudimentary democracies of the 19th century managed to accommodate slavery, they dispossessed and exterminated indigenous peoples in Australia and the Americas, among other places. In the 20th century, they have alienated minorities, accommodated anti-democratic parties (and on occasion, like Weimar Germany, been undermined by them).

And democracy is a lousy system because it's destined to muddle through, to survive by trading interest against interest, horse against horse. Politicians in democracies are forced to take the middle ground, to water down their proposals and policies. There's no exhilarating vision for us all to march towards. Apathy sets in. But, emergencies on planet earth – global warming style – notwithstanding, perhaps that isn't such a bad thing. Governments that are big on the 'vision thing' tend to build their success on the mass graves of those that disagree with them. During the 20th century alone, government actions in countries that were strong on ideology – the Soviet Union, China, Nazi Germany and Cambodia, for instance – cost some 100 million lives, independent of deaths caused in warfare, much of which was itself caused by ideological stridency. Self-inflicted massacres on such epic scales have been absent in democracies with their wishy-washy, middle-of-the-road politics and checks and balances on executive power.

Those MOR politics and checks and balances on power may look bland and insipid from the outside, and can be ugly and damned frustrating on the inside, but they are remarkably good at coming up with the kind of compromises and incremental progress that many political issues need. The question is whether the system is fit for Crunch Times when a more active approach is needed. The issue for us as fellow travellers through the 21st century will be how to maintain those checks and balances against the forces threatening to destroy them, from outside as well as in.

The domino theory in reverse

How can democracy be in jeopardy? Hasn't it all been going democracy's way in the last part of the 20th century? During the Cold War, the domino theory held that once one country succumbed to communism the rest would follow. History's tide, and the power and inevitability of Karl Marx's vision, meant that an irresistible red wave would roll in and submerge the weak, bourgeois democracies.

It didn't quite work out like that.

In fact it pretty much happened in reverse. Over the last quarter of the 20th century, one after the other – domino-style – authoritarian governments on the left and the right collapsed under the weight of their own internal inconsistencies. The process began in the mid-1970s in Southern Europe's top holiday destinations, with rightist military dictatorships in Portugal, Greece and Spain giving way to functioning democracies. In the 1980s, democratic governments arrived in South America. Peru, Argentina, Uruguay, Brazil, Paraguay, Chile and, in 1990, Nicaragua all replaced bullets with ballots. In Asia, the Philippines dumped Marcos in 1986, and South Korea voted for voting the following year. There were democratic reforms in Taiwan, and huge, if ultimately unsuccessful, democratic protests in China and Burma.

But the most dramatic switch was in Europe. Communism's flagship totalitarian regime, the Soviet Union, had begun struggling internally from the 1970s. By the mid-1980s Soviet leader Mikhail Gorbachev tried to stop the rot with his policies of perestroika (restructuring) and glasnost (openness). His country responded by falling apart. The Soviet Union found itself unable to maintain its hold on its East European empire, its acolytes began breaking away, and with the symbolic fall of the Berlin Wall dividing communist East Germany from democratic West Germany in 1989, the communist dominoes began falling. Although in most cases the transition to any recognisable form of democracy has been complex and troubled – Yugoslavia, for instance, split into pieces and tore itself to bits – at the beginning of the 21st century the threadbare flag of liberal democracy flies from Warsaw to Vladivostok.

The collapse of communism caught academics by surprise. Some even had to revise their lecture notes. This huge number of historical transitions couldn't be just coincidence! But what was going on? Into this intellectual vacuum stepped Francis Fukuyama, an American professor who rolled up these events and saw an opportunity to turn Marxism on its head and make a name for himself. In the summer of 1989, as communism lay dying, he wrote an article proclaiming the 'End of History'. Fukuyama argued that History (with a capital H) is the evolution of society to more advanced forms of civilisation – a process to which games maker Sid Meier no doubt subscribes. More than that, History has ended because, with communism calling it a day, liberal democracy is the only game in town. Period.

History – it ends with a ballot, not a whimper.

According to Fukuyama:

- As science makes breakthroughs, we simply know more about the world and evolve to a higher state of knowledge.
- Technology advances with science, and changes the way society is organised economically. As Marx said: 'The hand mill gives you society with the feudal lord; the steam mill, society with the industrial capitalist.' The only way to use today's technology efficiently is through advanced industrialisation and a market-based economy. Societies that don't will be trampled by those that do.
- But a market-based economy doesn't necessarily mean democracy. Think Singapore or China today, or Japan and Germany in the 19th century. BUT, a market-based economy does need better-educated workers.
- Man isn't just an economic animal – he needs other things apart from just food, shelter and sex. He needs *recognition* of himself as a human being with value and worthy of respect, and this need for recognition grows with his level of education. The only political system that can ultimately provide this recognition to all citizens is liberal democracy.

- So in the long run, democracy wins. History ends. Next!

Well, maybe. Market-based liberal democracy may yet turn out to be the end point of social evolution, but no one could deny that it has its problems, big ones. As a political system, democracy isn't even that peaceful and well ordered in those countries that have had it the longest. It produces its fair share of inequities, injustice and conflict, and, as we'll see later in the chapter, its legitimacy is constantly being threatened by real and significant structural problems. So history with a small h continues, and the crunch for us in democratic countries is how we organise ourselves to respond to the threats to our least-worst political systems from both the outside and the inside. What are these threats?

Manipulating democracy today

Men like Lippmann and Bernays, originally working for Woodrow Wilson (one of America's most liberal, do-gooding presidents, aiming to make the world safe for democracy), developed the black arts that threaten democracy today. In the Crunch Time world, those black arts combine frighteningly with the increasing dominance of the mass media.

This nightmare was foreseen many decades ago, predicted by a left-wing writer who invented a world in which the media would dominate our daily lives, filling our minds with garbage that crowded out anything of significance. The writer was George Orwell, the book *1984*. Here's Orwell writing about the Ministry of Truth, the propaganda arm of a vicious totalitarian regime manufacturing news for the 'proles': 'Here were produced rubbishy newspapers containing almost nothing except sport, crime and astrology ...' Remind you of any newspapers you've read lately?

Orwell was right: media tycoons like Rupert Murdoch and Silvio Berlusconi have made fortunes keeping the masses contented with *1984*'s ironic recipe of sport, crime and astrology (Orwell thought they'd keep sex for the movies – you can't be right about everything).

Murdoch and Berlusconi have both used their empires to further political agendas, but from opposite ends of the spectrum. Although Murdoch is a US citizen he has switched countries, and some might say wives, in furtherance of his international business interests. Berlusconi is strictly Italian and used his media business to secure the leadership of his country. The average Italian soaks up more than four hours of TV a day, and Berlusconi provided their diet of sex, sport, crime and astrology with his three private Mediaset channels.

How was a man with so much media power allowed to hold so much political power? He promised to change the law so that when he became Prime Minister his media interests would be quarantined in a way that would put them out of his reach. It's difficult enough to imagine how this might have happened had he been true to his word. Needless to say, he wasn't.

That isn't the worst of it. When he became Prime Minister in 2001, Berlusconi faced four criminal charges, including false accounting, illegal financing of a political party, and bribing judges. One by one, he managed to wriggle out of facing them, through legal wrangling, refusing to cooperate or simply changing the law in his own favour.

Italian journalists grew used to self-censorship, fearing they might lose their jobs if they were overcritical of Italy's richest man, their Prime Minister. Cases of 'improving' the news ranged from editing out prime-ministerial gaffes to airbrushing his bald patch on the cover of *Panorama*, a weekly magazine.

But remember, Italy is a democracy. People did actually have to go to the ballot and elect him. And they could vote him and his party out – and in 2006 they did. Just. You can't buy people in a democracy. Or can you?

Bernays and Lippmann put the lie to that innocent notion a long time ago. Money buys voice, and printing more election leaflets than the next guy pales into significance when it comes to owning your own TV networks, newspapers and publishing companies. Did we forget to mention Berlusconi owns Italy's biggest publisher with 30 per cent of the book market, that he controls 38 per cent of the magazine market, and owns the country's largest newspaper, *Il Giornale*? In his successful 2001 election campaign, he actually sent out 12 million free copies of his autobiography.

Italy isn't the only democracy suffering from the effects of too much money in too few hands. In the US 1999/2000 election cycle some $3 billion was spent to elect federal candidates. The politicians elected by these tidal waves of cash aren't just in hock to their donors. They're also pandered to by lobbyists with their own formidable financial resources. Politics and money have always been bedfellows, just as Church and state used to be. It took several centuries for that link to be broken in much of the Western world, although when you look at the support George W. Bush receives from and gives to the fundamentalist arm of the Church in the US you have to wonder …

not to mention the Church tax that's still collected by the state in Germany. The question is whether it will take 200 years for campaign finance reform to sweep the democracies of the world. Let's face it, we don't have that long: it is, after all, Crunch Time.

So, is there anything that might save us from being bought? Some people think there is.

Education, education, education

Learning about democracy in school is like reading holiday brochures in prison.

Derry Hannam, British educator

Super-cynic Walter Lippmann reckoned that society, with all its creases, folds and facets, is too complex for voters to understand the real implications of their political choices. Victorian thinker John Stuart Mill reckoned the necessary precondition of universal suffrage was compulsory secondary education.

And today? Should a working knowledge of statistics be required? Or an exam for potential voters, like a driving test? Should your vote, as a person intelligent and curious enough to get at least halfway through a complex book such as this one (or be lucky enough to open it at such an interesting page) be worth more than someone whose biggest effort at reading is the label on their beer can?

When it comes to education the democratic waters get pretty muddy. Political philosophers have been hung up about education since ancient Greece. The idea of education in a democracy is to make us all think differently. Education, argue the theorists, produces informed disagreement, however bitter. Without it we're simply flying without instruments. American thinker John Dewey wrote *Democracy and Education* at around the time Walter Lippmann was rallying Americans to join the Western Front. Dewey thought education was the glue that held democracy together, a shared experience that would help us all appreciate one another's points of view. Perhaps he was overstating the case, but as we shall see in Chapter 10, and as we saw in Chapter 7, there's no doubt that the more we understand about the world, the more crucial our democratic institutions become.

Even so, Lippmann knew that education may give us more understanding, but it doesn't prevent us from following our irrational, animal instincts (even world-famous Cambridge physicist Stephen Hawking gets photographed in lap-dancing joints). We're easily lured into endorsing decisions by slogans, ads, posters, sound bites, TV spots – we buy wars as willingly as we buy trainers. Is democracy just a fancy name for the system that vested interests use to manipulate us?

Democracy's Disneyland

On the shore of Lake Zurich sits the municipality of Kilchberg, a busy, innocent little place, home to about 7,400 people. It's the headquarters of Lindt chocolate, but unfortunately, chocolate has little to do with the future of democracy. What does is what goes on in Kilchberg itself, a model that perhaps all democracies could work towards, a method of local government that encourages and rewards citizenship and community involvement, and is also transparent and democratically accountable.

The municipality itself holds all the powers that haven't been specifically given over to the regional or federal government. A quarter of all the taxes people pay go to the municipality itself, through income and property levies. Kilchberg educates its children up to the age of sixteen, and does everything from build the school to elect the committee that hires the teachers. It hands out cash to its poor and to a handful of refugees. It has a volunteer fire brigade with a youth section. The local police run a couple of patrol cars in town and a couple of boats out on Lake Zurich. There's an old people's home and a community farm with an honesty box. And seven elected councillors – that's one to every thousand or so inhabitants – who supervise a small team of professionals, run it all.

But that's not why little Swiss municipalities are so great. The real kicker is that every three or four months at eight o'clock in the parish hall the seven people who run Kilchberg have to present their recommendations to an open meeting and let the voters decide on what to do. These meetings fix taxes, pass new laws, check the accounts and OK planning proposals. Anything else anybody wants to bring up can be discussed. This is direct democracy. Want to vote? Raise your hand. Want a paper ballot? No problem, but a third of people must agree. If you don't like the council's ideas, go get fifteen signatures and put a new proposal to your fellow voters. A single person can demand some specific other action from the council, with the right, if the council doesn't agree, to take the matter up to canton and federal level. In 2005, the people of Kilchberg voted on thirteen different local issues, and turnout ranged from 40 per cent to 67 per cent.

Before you roll your eyes at this cheesier-than-Swiss-cheese story, you would do well to note that the world's only superpower is also built of little communities just like this one. Even if you don't agree with how the US behaves in the international arena, there's something commendable about the way its citizens involve themselves in their own governance. In America, town meetings are used to elect almost all municipal officials, from the town clerk and sexton through to the school-district treasurer and the head of the Parents and Teachers Association. A century and a half ago Alexis de Tocqueville, travelling French aristocrat, noted in *Democracy in America* that: 'Town meetings are to liberty what primary schools are to science; they bring it within people's reach, they teach men how to use and enjoy it. A nation may establish a free government, but without municipal institutions it cannot have the spirit of liberty.'

It's that spirit of liberty that's missing when we feel disempowered and disillusioned by the weakness of our vote. In Crunch Time when global forces affect our communities and families, we compound the problem by giving away control of the parts of our own lives that should rightfully be ours. We become disengaged from the only communities we have direct influence over – our own. The way forward is to take them back, to actively inject life back into our local communities, to access the energy that has been stolen from us by the everyday demands of being a person in the 21st century and reinject that through democracy into other people and other communities.

Democracy means that power is vested in the people, and should be given upwards only where necessary. It's something the Swiss are big on.

Globalisation. Again …

But there's something the Swiss aren't big on. And that's being big. Switzerland doesn't have the people, the military or the economic clout necessary to get involved in global politics. Unlike the Americans, who are big on being big, the Swiss keep themselves to themselves. In global affairs, they are hapless spectators.

Take the ozone layer for example. When countries got together in 1989 to ban the chemicals that were destroying the ozone layer, no one realised that their replacements would speed up global warming. The Swiss did. They tried to sound the alarm in 1990. The reaction: nothing happened for sixteen years. Only now are meetings to talk about the problem even taking place. Meanwhile the glaciers on the Eiger continue to melt.

Crunch Time issues are simply beyond the reach of many of our democratically elected governments. With globalisation, our own actions reach across borders in a way that makes democratic governance irrelevant. We buy timber, diamonds or oil which prop up corrupt regimes abroad, or bankroll lobbyists for big corporations in our own backyard.

As the crowds of protesters outside the G12 meetings will only too willingly attest, globalisation poses a huge threat to democratic accountability. Why? Because we vote within the structures of a nation state, while global issues batter away at our ability to control what goes on without so much as a democratic nod or whistle. A financial scandal in Houston hits Arthur Andersen employees in major cities around the world. Our democratically elected representatives can tantrum about it all they like, but nurses, teachers, assembly plants and call centres are shipped from continent to continent, leaving us back home without health, education or jobs. The bits of our lives controlled by our democratic institutions, be they parliament or congress, senate or Bundestag, are shrinking.

Countries that don't adopt the neo-liberal agenda in their economics are cut out of the wash of global funds. What politician can afford to promote a policy that will cut his country off from the global financial system and the billions of dollars of foreign investment it offers? None that expects to remain in office. In globalisation circles, this has become known as the 'golden straitjacket'. If you wear it the right way (the neo-liberal way), it, apparently, comes with great financial rewards. But once you've got it on, freedom of movement is, well ... restricted. Whether it makes countries richer or simply more volatile, the golden straitjacket and the policies it implies undeniably cuts down the range of policies our governments are able to propose or implement.

If you are feeling disempowered by the political process, it's because politicians long ago admitted defeat in the face of the seemingly overwhelming power of globalisation. Down under, it began with Bob Hawke, as the one-time trade unionist held the white flag to the global financial markets and implemented a string of liberalising and deregulating reforms which continued through the Keating and Howard years, from floating the Aussie dollar and striking a wages accord to deregulating the banks and implementing a goods and services tax. The country might be financially better off as a result (and it might not) but who can argue that left-winger Hawke and his successor would have followed these policies if right-wing Thatcher and Reagan hadn't.

The world beyond the borders of our own countries often forces its way into our lives. War, trade, diplomacy; these things have always materially changed people's circumstances. Think of the British troops who are today fighting to keep the peace in Basra, a provincial town in southern Iraq, or the hundreds of thousands who fought in the world wars. But in Crunch Time, it's not just commercial connections that count, it's the things that people do within their own borders that affect us materially – how much carbon dioxide America spews into the atmosphere matters to you and I; whether the Chinese protect intellectual property matters to British inventors.

The influences that touch our lives are global, but our politics remain (in)effectively local. The old feudal aristocracy was tied to land. Dukes and counts were always of somewhere: a piece of land, a place on the map. Feudalism may have disappeared but democracy – by necessity, almost by definition – keeps its love of land alive. Our representatives remain resolutely tied to geography, which limits their ability to respond to the global influences that increasingly shape our lives.

But this presents us with an immense paradox. We need strong international institutions to solve the big global problems Crunch Time is presenting us with: global warming, the seepage of nuclear weaponry, an international water crisis and the rest. But the creation of such institutions implies a loss of the democratic accountability that's crucial to maintaining healthy societies. Even now, at the beginning of the 21st century, much of the power our democracies gave to national parliaments lies in some foreign city with some institution we only know as some obscure acronym: EU, UN, APEC, WTO ... the list goes on. The power hasn't evaporated, but the democratic accountability has.

According to democracy advocates, whenever our leaders sign up for a big international agreement they give up a little of the power that we gave up to them. Today that means agreements not just on old-fashioned things like military alliances, it means environment protocols, farm subsidies, accounting standards – you name it, they've signed up to it. The voters have been squeezed out again.

What does this mean for us? It means that broad, global Crunch Time issues will be the source of increasing friction through the next century.

Democracy's flexibility

Of course we live different lives now than we did when de Tocqueville was visiting America. Frankly there's hardly enough time to do the things you need to do, let alone attend a rambling and time-consuming meeting about the status of the municipal toilets. Even in Kilchberg, no more than about 400 people generally turn up at town meetings, maybe 700 when something especially exciting is on the menu. That's only a small percentage of the community's 4,000 qualified voters.

The answer lies in two things: the advent of new technology, particularly the all-pervasive internet, and new forms of democratic governance that could help us address some of the broader issues we've identified.

Firstly, instead of politicians pushing our buttons perhaps we should be pushing theirs. For the past 200 years – except in the odd Swiss canton, and in American country towns – democracy has meant a system by which people vote every few years to elect a handful of representatives, who in between elections take all the important decisions: war (think invasion of Iraq), peace, taxes, the lot. Technology now offers us the opportunity for something more direct, more fully 'democratic' – decisions by vote of the whole people. That ought to set Walter Lippmann rolling in his grave.

As he once noted, the practical justification for representative democracy was that people needed their betters to represent them. Now representatives say it's because of a separation of roles. We, the people, are actually too busy to govern the country (although a look at empty seats in assembly chambers will show that often our representatives are too). But during the last half of the 20th century, most people in the Western democracies have got themselves better educations, more money, and more free time to think about what goes on around them. They're better informed too, and are used to discussing issues as broad and varied as the impact of genetically modified foods on the environment, the justification for the invasion of Iraq, or the plots and sub-plots of *Lost* episode by episode.

In time, people will grow to expect more from their democratic system than that which is being delivered and challenge the structures of representative democracy. The internet will eventually provide reliable methods for validating electronic votes, and will remove the biggest single obstacle to direct democracy – the physical difficulty of distributing information to a large population, engaging it in debate and collecting its votes. In light of these developments, many people will come to see national elections every few years as a totally blunt instrument for expressing the popular will, a remnant from the age of steam, when most representative institutions were invented.

Just ask

Pollsters regularly ask our opinion on issues of state, and elected representatives pay attention to what the opinion polls say. Everywhere, ordinary people are now in a better position to examine what their representatives are up to, observe their voting records, web pages or personal problems. They can conclude for themselves whether it's really a good idea to let this pathetic shower choose so many of our policies. The politicians seem to be making the same choices too. People are no longer willing to offer the deference their representatives used to expect.

What's more, Swiss-style direct democracy may be better than the representative sort at coping with one of the chief weaknesses of modern politics – lobbying. In the relatively humdrum, de-ideologised politics of post-communist days, the lobbyist is getting even more powerful than he used to be; and democrats are right to be worried.

Lobbying has an important role to play in policymaking. People who take decisions, in any field, should be party to as much argument and debate as possible. But Lippmann and Bernays taught us how persuasive persuasion can be. Lobbying goes wrong when special interests use their money to cross the line between persuading politicians and pocketing them. In dealing with a relatively small handful of elected politicians, the lobbyist has many ways of doing that, ranging from 'entertainment' through to the manila envelope stuffed with cash, or the legal donation or loan of cash into campaign funds. But when lobbyists face an entire electorate, bribery and vote buying are more difficult. Advertising billboards and 30-second slots replace cosy chats in clubby corners.

Wealthy media tycoons can help voters make dumb decisions. For the big issues with big interests involved, this will always be so. However, the lesson of Crunch Time is that we need to take control where we can, to drag the power back down to a level where we can exert influence over the forces that affect our lives. Yes, we will make mistakes, but at least they will be our mistakes, not those of some infuriatingly unseen bureaucrat or smug politician.

The more political responsibility ordinary people are given, the more responsibly most of them will vote, goes the mantra. Direct democracy, with a little Crunch Time technology, may help to produce something closer to true government by the people. And that, after all, is the way the logic of Crunch Time points. Democracy's cheerleaders and ideologists spent much of the last century telling totalitarians and dictators that they ought to trust the people. Now it's time for them to trust us.

CHAPTER 10

Extreme Evolution

How science may help us out-think the future

Isaac Asimov's three laws of technology:

1. When a Scientist states that something is possible, he is almost certainly right. When he states that something is impossible, he is very probably wrong.
2. The only way of discovering the limits of the possible is to venture a little way past them into the impossible.
3. Any sufficiently advanced technology is indistinguishable from magic.

Here in spades is the incurable optimism of the science-fiction writer. Real scientists are a little more nervous about our prospects. Wheelchair-bound physicist Stephen Hawking believes that humanity has just a few decades to escape the planet. How's that for Crunch Time.

✪

Hundreds of thousands of words. Thousands of emails. Thousands of hours on the internet. Hundreds of hours on the phone.

The book you're reading is more than just print on page. It's the outcome of a dialogue that we constructed over time and distance. Mike, from his backyard shed in Bondi, sends the final fruits of his day's research and writing to Adrian, who picks it up wirelessly on the commute from Kent to London's St John's Street. From there, it's sent home to Kent, worked on and resent back to the shed in Bondi for Mike to pick up and work on in the morning, Sydney time. This book would've been all but impossible if the manuscripts had been sent by mail. The same is true of literally billions of human endeavours across the globe today: modern life is defined, shaped and made possible by the magic that science has uncovered for us.

The products of scientific and technological innovation frame our every waking moment. Call it what you will, the internet era, the space age, post-industrial society, Crunch Time, we're living in an age of magic, and by most traditional measures are much better off for it.

Still, rushing round from appointment to appointment, picking up text messages and popping our selective serotonin reuptake inhibitor, we're beginning to wonder about where the balance of power lies between ourselves and the science we use to live our lives.

There's a niggling feeling that science and its fruits are conspiring against us, not-so-subtly trapping us in a cycle of invention and obsolescence, taking advantage of our human curiosity, needs and desires for some end we're not clever enough to see. Unstoppably, science itself chips ever closer to truths that we don't necessarily want to know.

Whether you like it or not, 21st-century science and technology has a distinctly Crunch Time hue. It's handing us the tools of our own destruction, while providing people with a never-ending series of distractions to while away the hours before our extinction. Sound overdramatic? Read on.

How could they do it?

> *From the violence of that salt ... so horrible a sound is made by the bursting of a thing so small, no more than a bit or parchment containing it, that we find the ear assaulted by a noise exceeding the roar of strong thunder, and a flash brighter than the most brilliant lightning.*
>
> Roger Bacon on gunpowder, *De secretis operibus artis et naturæ*

> *I have felt it myself. The glitter of nuclear weapons. It's irresistible if you come to them as a scientist. To feel it's there in your hands, to release this energy that fuels the stars, to let it do your bidding. To perform these miracles, to lift a million tons of rock into the sky. It's something that gives people an illusion of illimitable power, and, in some ways, it's responsible for all our troubles – this, what you might call technical arrogance, that overcomes people when they see what they can do with their minds.*
>
> Freeman Dyson on the atomic bomb

On 16 July 1945, Robert Oppenheimer stood on the edge of the New Mexico desert and watched the world's first atomic bomb explode.

Oppenheimer led the Manhattan Project, the team that produced the first nuclear weapon. This was science on a massive scale: 50,000 people spent four years of their lives creating that explosion. The money spent could have built more than 3,000 B-29 Superfortress bombers, the aircraft that eventually carried the atomic bomb to Hiroshima.

Looking back on that time, as the threat and consequences of the proliferation of nuclear weapons fills our newspapers and our nightmares, one question dominates: how on earth did the Manhattan Project scientists – an extraordinarily moral and self-critical group – allow themselves to build the most destructive weapon in history, one that could literally threaten the continued existence of mankind? The question is worth considering because it runs to the heart of our understanding of how science progresses, its future in these complex and uncertain times, and the direction and application of technology.

Many of the scientists had their own strong ethical views on the use of the bomb, but they carried little weight for the politicians and soldiers fighting the Second World War. The sheer horror that conventional weapons had wrought over years of conflict compromised countless moral arguments in the name of expediency. At first, objections to the bomb were based on its use against civilians. When the war had begun in 1939, a neutral United States had warned both Britain and Germany against bombing cities. By the time of Oppenheimer's test, bombing civilians had become routine for all sides. The Luftwaffe's Blitz had given way to Allied air attacks on cities like Dresden. In the Pacific theatre in March 1945 more than 300 United States Superfortress bombers conducted a massive incendiary raid on Tokyo. The bombs they dropped razed sixteen square miles of the Japanese capital and killed as many as 100,000 people.

When the first operational atomic bomb was ready to be used, the moral argument had evaporated. The only material difference was that the concentration of power had changed – a single bomber could now do the work of hundreds. The bomb dropped on Hiroshima destroyed five square miles of the port and killed 40,000 civilians. Its scale of destruction was small when measured against the Tokyo raid's still more appalling devastation.

The scale of the intellectual and technological challenge and the unfathomable complexity of the project allowed the scientists responsible for taking civilisation over the nuclear threshold to simply shift the blame. A simple compartmentalising of responsibility solved their ethical dilemmas. In a memo recommending the immediate use of the bomb on a civilian target in Japan, Oppenheimer wrote that scientists have 'no claim to special competence in solving the political, social, and military problems which are presented by the advent of atomic power'. Perhaps these words of symbolic hand washing, buried in a document intended for a readership only at the highest and most secretive echelons of the United States administration, mark the moment when modern science turns to us and says: 'You're on your own.'

Three years later, after Hiroshima and Nagasaki and with Japan defeated and the euphoria of discovery spent, Oppenheimer reflected rather differently on his moral obligations. 'In some sort of crude sense which no vulgarity, no humour, no overstatement can quite extinguish', he wrote, 'the physicists have known sin; and this is a knowledge they cannot lose.'

In 1947 the *Bulletin of the Atomic Scientists*, founded by Oppenheimer, Einstein and other intellectual leaders of the time, began putting what it called the Doomsday Clock on its cover, out of concern about the impact of the science they had helped to create. Ever since, it has shown the scientists' estimate of the danger of complete annihilation, reflecting changing international geopolitics. At the turn of the century, the hands on the clock had moved seventeen times in its history, swinging backwards and forwards with the signing and breaking of non-proliferation agreements, the outbreak of war in different parts of the world, and the increasing and decreasing of tensions in various strategic theatres. On 27 February 2002, the *Bulletin*'s board of directors moved the clock forwards by two minutes, from nine to seven minutes, the same setting that the clock hands had been placed at 55 years before, at the very beginning of the Cold War.

On 18 January 2007, at a joint press conference held at the American Association for the Advancement of Science in Washington, DC, and the Royal Society in London, the scientists moved the clock forward by two minutes to five minutes to midnight. Physics genius Stephen Hawking said: 'A group of scientists in 1945 took responsibility for the consequences of the atomic bomb they had helped to create ... since Hiroshima and Nagasaki, no nuclear weapons have been used in war, though the world has been uncomfortably close on more than one occasion. But for good luck we would all be dead ...'

The second nuclear age is unfolding, say the scientists. 'We stand at the brink of a Second Nuclear Age', announced the *Bulletin*. 'Not since the first atomic bombs were dropped on Hiroshima and Nagasaki has the world faced such perilous choices. North Korea's recent test of a nuclear weapon, Iran's nuclear ambitions, a renewed emphasis on the military utility of nuclear weapons, the failure to adequately secure nuclear materials, and the continued presence of some 26,000 nuclear weapons in the United Sates and Russia are symptomatic of a failure to solve the problems posed by the most destructive technology on Earth.'

The scientists remind us that the threat of nuclear annihilation hasn't receded since the weapons were created, even if the passage of time has dulled our fear of 'the bomb': even the word sounds a little old-fashioned. This time, however, the *Bulletin* added climate change, genetic engineering, and other Crunch Time perils to the list of threats to human civilisation.

The scientific architects of our potential destruction are no longer limited to the world of nuclear physics. Different spectres of extinction have risen up in parallel. While the image of the mushroom cloud suggests that humanity's demise will come with one big bang, the kind of science that hits the headlines these days sees our decline more in terms of an evolutionary whimper.

New threats from science

Scientific knowledge grows exponentially. That means discoveries and the technologies they create happen faster and faster. And that means potential threats emerge faster too.

This manifests itself in many ways. As just one example, take the ability of manufactured machines to compute stuff. It has been growing exponentially for the last hundred years, and will more than likely continue to do so for as far into the future as we can see. Technology buffs refer a lot to something called Moore's Law on Integrated Circuits. Its author is Gordon Moore who helped invent integrated circuits and started chip-making giant Intel. In 1965, six years before Intel rolled out its first microprocessor, Moore noted that the surface area of a transistor (as etched onto an integrated circuit) was being reduced by a half every year. A decade later, he revised this to every two years – which is, apparently, a better fit to the data.

It's not a law, just an observation that has worked up to now. Every couple of years we get twice as much circuitry running at twice the speed for the same price. This has been going on for as long as the computer has been around, and the result is the wonderful devices we use today for everything from email or writing our last will and testament to monitoring the state of shuttle launches and the productivity of toilet-tissue factories.

But the fact is that the exponential growth in computing power has been going on for much longer than the integrated circuit has been around. Ray Kurzweil, a leading international expert on artificial intelligence and the inventor of the world's first print-to-speech reading machine for the blind, claims to have discovered the Exponential Law of Computing. Kurzweil tracked back to the first use of computing technology in the 1890 United States census, through relay-based computers that cracked the Nazi's Enigma codes, to vacuum-tube computers of the 1950s, to the transistor machines of the 1960s, and to all of the generations of integrated circuits of the past four decades. 'Computers today', says Kurzweil, 'are about one hundred million times more powerful for the same unit cost than they were a half a century ago'.

And this progress will continue. Some say that the limits of silicon will be reached within the next twenty years. Evidence and intuition suggests that something (nanotechnology, perhaps) will enter to fill the void, and enable yet more improvements in computer processing power indefinitely into the future. By the year 2010, there will be 10 billion transistors on the face of the earth – per human being. The power of each of these is multiplied because they work together – exponential rather than arithmetic development means 20,000 years of computing 'progress' will be crammed into the next 100.

We all recognise the impact computers have had on our lives today, but what are its implications for the century ahead? While you consider this, remember that the ability of human-made machines to compute stuff is just one avenue of scientific endeavour. Computing progress enables forward leaps in genetics, robotics, nanotechnology, even archaeology. But before we get too excited about a genetically purified, wire-free networked world, what terrifies some commentators is the way that these technologies combine, multiply and reinvent themselves. Each one has the capability to outstrip the ability of governments, universities and even corporations to control them.

The chief whistle-blower here was Bill Joy, top scientist at Sun Microsystems and an unlikely anti-technologist. Way back at the turn of the millennium, Joy wrote a long and compelling article in techie magazine *Wired*, outlining his own fears for the future of humanity, threatened by the power of the new sciences in which he and his colleagues were working. The article was designed to kick-start the debate regarding the management and control of 21st-century technologies. It certainly did that.

The debate around technological development and its ability to outflank humanity's own evolutionary progress focuses on four particular technologies, genetics, robotics, information and nanotechnology – known as GRIN. The reason why these technologies are more Crunch Time than earlier ones is because they can all potentially self-replicate. 'A bomb is blown up only once', says Joy, 'but one bot [robot] can become many, and quickly get out of control'. Some quick definitions:

Genetics: The study of heredity and inherited characteristics. Although the science of genes presents almost unlimited potential to aid in the treatment of disease and improve health, commentators worry that its potential to tinker with human nature itself will be open to abuse from our less noble traits: vanity, prejudice and the like. Worriers' alarms are set ringing by the very real possibility of genetically engineered plagues, viruses and the like, which when released into the population or the environment by malevolence or incompetence will prove uncontrollable.

Robotics: The search for the ultimate robot demands a deep exploration of what it means to be intelligent and sentient, in fact what it means to be human itself. The ultimate fear is that by delving too deeply into the human condition we'll discover that in actual fact we're no more than a bunch of wires connected by electrical impulses – just a sophisticated robot. Some say that controlling sentient, self-replicating robots with superhuman physical and mental powers will be beyond us.

Information: As computing power and our ability to use it skyrockets, so does our dependence on the information held in the system. If your computer crashes, that's it for your working day (remember the kerfuffle over the Y2K bug?). As society comes to depend on a network of incomprehensibly complicated databases and connections, so it becomes reliant on it working properly. Enter the bad guys, or simply complexity beyond our competence.

Nanotechnology: Seeks to build objects at the molecular level, by arranging atoms in a particular way.

> Scientists envisage the ability to build everyday objects molecule by molecule by specifying the atomic characteristics of the object and sitting back while a nanotechnology machine goes to work. The out-of-control scenario foresees self-replicating, robotic, genetically engineered nano-replicators that begin to eat everything, consuming the world's atomic material and destroying all organic life.

Picking up where Joy left off, Martin Rees, Britain's Astronomer Royal, in his book *Our Final Hour: A Scientist's Warning*, lists a cornucopia of catastrophes waiting to happen. On top of terrorism, smallpox and global warming, we now have to worry about microscopic self-replicating machines the size of molecules, reproducing out of control; lethally engineered super-pathogens creating unprecedented pandemics that wipe out great chunks of humankind; and particle-accelerator experiments that create planet-destroying black holes.

Back in the 1940s before lighting the fuse at Trinity, Robert Oppenheimer was worried that an atomic explosion might set fire to the atmosphere in an uncontrolled chain reaction. Oppenheimer was so troubled that he consulted his mentor, Arthur Compton, who suggested a risk/benefit calculation. He reckoned it would be better to lose the war with the Nazis than risk destroying the earth's atmosphere, if that risk were greater than three in a million. Rees's estimate of the likelihood of science destroying mankind in the next hundred years is much greater. 'What happens here on Earth, in this century, could conceivably make the difference between a near eternity filled with ever more complex and subtle forms of life and one filled with nothing but base matter.' Crunch Time again.

But we might not end up in Rees's pieces. His theories are simply the latest contributions to a debate that has been seesawing since the Manhattan Project gave us the ability to unleash self-inflicted planetary destruction. They just took an up-tick in volume and urgency as the 20th century ended.

Party like its 1999

At birth the infant will be clamped in front of the TV eye by means of a suitable supporting structure, and two sections of tubing will be connected to provide nourishment and to carry away the waste materials. From this time on, the subject will live an ideal vicarious life, scientifically selected for compatibility with the fixed influences of the inherited genes and chromosomes.

Daniel Noble for Motorola, 1962

Put it down to pre-millennial tension if you like, but 1999 was a great year for the philosophy of science in popular culture. In that year Ray Kurzweil published *The Age of Spiritual Machines*, and another influential computer scientist Hans Moravec came out with *Robot: Mere Machine to Transcendent Mind*, which proclaimed that in the coming century our computational creations will outstrip us intellectually and spiritually. Perhaps more than anything else, these two books illustrate a slightly weird and misanthropic tendency in computer scientists: the scenarios they paint suggest a little too much Star Trek watching. But they're also the imaginings of men who have a serious grasp of the possibilities of current technology and so demand serious consideration.

Kurzweil's and Moravec's ideas were eerily realised in

The Matrix, a movie that came out in the same year featuring spectacular screen effects, sci-fi and martial arts backed up by a blank-featured Keanu Reeves as a computer hacker given some startling news. Reeves's character, Neo, is told that contrary to the impression he's given by everything in the world around him, it's not 1999. It's actually a couple of hundred years later. The earth we know has been destroyed in a war between man and the machines he created. We lost. But rather than destroying us all and turning to wind farms for energy, the vindictive machines have placed humans in an orderly pod farm – an enormous bioelectric power plant – in which we're nurtured from birth until death, as kind of biological Duracell bunnies. But (in an unexplained moment of generosity) instead of keeping us comatose, they create a true-to-life world for our minds: the Matrix, an artificial world as 'real' as the one we see around ourselves today, injected directly into our minds through a coaxial jack in the back of our heads. When Albert Einstein said: 'Reality is merely an illusion, albeit a very persistent one', *The Matrix* probably wasn't what he had in mind.

The Matrix – daft plot aside – translates a very Crunch Time human fear: that we may be superseded in the evolutionary race by the very devices we invented to assist us in it. Is this really possible? For Kurzweil, Moravec and their chums, it's not only possible, it's almost inevitable, and is happening already. Not in a Matrix-like confrontation between men and machines, but through stealth. Kurzweil even has a timeline.

2009 A basic PC can perform a trillion calculations per second. Web access is high-speed and wireless. Most routine business transactions (purchases, travel, reservations) take place between a human and a virtual personality, which often includes an animated visual presence that looks like a human face. Bioengineered treatments for cancer and heart disease have greatly reduced mortality from these diseases. A neo-Luddite movement is growing.

2019 A basic PC can now equal the computational ability of the human brain. Computers are embedded in walls, tables, chairs, desks, clothing, jewellery and bodies. 3-D virtual reality displays, embedded in glasses and contact lenses, as well as auditory 'lenses', replace mobile phones. Nano-machines are beginning to be applied to manufacturing and process-control applications. Automated driving systems are installed on roads. People are beginning to have relationships with automated personalities and use them as companions, teachers, even lovers.

2029 A basic PC has the computing capacity of 1,000 human brains. Permanent or removable brain implants now provide inputs and outputs between humans and the 'grid'. Direct neural pathways have been perfected for high-bandwidth connection to the human brain. A range of neural implants is becoming available to enhance visual and auditory perception and interpretation, memory and reasoning. Humans now communicate and interact with machines more frequently than with people.

2099 Geek-future ends when '... there is no longer any distinction between humans and computers ... Most conscious entities do not have a permanent physical

> presence ... Life expectancy is no longer a viable term in relation to human beings ...'

To Ray Kurzweil, and thinkers like him, the redundancy of humanity doesn't seem unlikely or repugnant. Others see it as a foregone conclusion. Nick Bostrom, a philosophy don at Oxford University, has published several papers that attempt to demonstrate mathematically that the probability that we're living in a Matrix-type computer simulation is much greater than the probability that we're not. All you really need to do to come to this conclusion is assume that computing technology continues to progress at the exponential rate we've been seeing over the last decades, and that the human race doesn't meet extinction (through, say nuclear or environmental holocaust) before we reach the post-human future. His cheerful conviction that the future is robotic led him to found the Transhumanist Association. This is their declaration:

> (1) Humanity will be radically changed by technology in the future. We foresee the feasibility of redesigning the human condition, including such parameters as the inevitability of ageing, limitations on human and artificial intellects, unchosen psychology, suffering, and our confinement to the planet earth.
>
> (2) Systematic research should be put into understanding these coming developments and their long-term consequences.
>
> (3) Transhumanists think that by being generally open and embracing of new technology we have a better chance of turning it to our advantage than if we try to ban or prohibit it.

(4) Transhumanists advocate the moral right for those who so wish to use technology to extend their mental and physical (including reproductive) capacities and to improve their control over their own lives. We seek personal growth beyond our current biological limitations.

(5) In planning for the future, it is mandatory to take into account the prospect of dramatic progress in technological capabilities. It would be tragic if the potential benefits failed to materialize because of technophobia and unnecessary prohibitions. On the other hand, it would also be tragic if intelligent life went extinct because of some disaster or war involving advanced technologies.

(6) We need to create forums where people can rationally debate what needs to be done, and a social order where responsible decisions can be implemented.

(7) Transhumanism advocates the well-being of all sentience (whether in artificial intellects, humans, posthumans, or non-human animals) and encompasses many principles of modern humanism. Transhumanism does not support any particular party, politician or political platform.

Great chess players, lousy winners

Putting Bostrom's boffins aside, there's real reason to be sceptical that computers will ever come close to being able to exist with anything near the richness of the dullest human life. Even the great milestone for the artificial intelligence geeks – the defeat of chess grand master Garry Kasparov by IBM's computer Deep Blue – was something of a damp squib.

When Kasparov was beaten IBM's share price soared. Nerd philosophers loudly proclaimed the battle between human intelligence and computer intelligence had been won – by machines. IBM hedged their bets, and dismantled Deep Blue the day after, thereby scotching any chances of a human rematch (and also preventing it from taking over the world).

But IBM's chess monster was programmed, built and run by humans to play a very human game. Kasparov's response to defeat was equally human. 'At least I have feelings about losing', he said. Kasparov went on to propose that chess be continued with players working with computers to augment the game. His response cuts to the chase. Chess is one of the ways that people fill their days. For some it's a living. But game it is, and game it will remain. Computers are unlikely to derive any satisfaction from pairing off to play a game of chess.

Machine 'intelligence' – even in the strictly defined world of chess – isn't much more than that of a sophisticated pocket calculator. Deep Blue calculated 200 million positions per second, while Kasparov generally managed three or four. But the point here is, humans don't need more than three or four moves to see patterns, to look into the future, or to make judgement calls, good or bad. Deep Blue needed 200 million moves a second just to win a chess match. How many would it need to save a bad marriage, sail a yacht or cross a busy street?

Five years after Kasparov lost, one of his pupils, Vladimir Kramnik, took on a more advanced computer – Deep Fritz. He drew in an epic contest, and in fact, he came close to winning. Artificial intelligence experts admitted that Kramnik had played the more elegant chess, in other words, lessons had been learned. When Kasparov faced Deep Blue, he played the computer like a human adversary – selecting complex positions and trying to out-calculate his opponent. Deep Blue just out-crunched him. Kramnik, by contrast, dramatically simplified his matches with Deep Fritz by removing big pieces, like queens, in early forced swaps. Three games in to the eight and Kramnik led by two and a half points to a half. But the computer didn't tire and clawed its way back against an exhausted opponent, who blew the fifth game to end up drawing the series.

Chess is a rule-based game, and computers have changed those rules. When Garry Kasparov started learning the game he relied on books of openings. Now anyone learning chess can consult an online database of nearly three million grandmaster games and statistics on the best moves. The game has become harder, but its qualities remain the same. And computers? In the words of the man who refereed the contest between Kramnik and Deep Fritz, himself an expert on artificial intelligence, computers remain 'pocket calculators'.

Dealing with Crunch Time science

So technology advances, wheedles its way into our lives, and we feel threatened and insecure, confused about where we're taking ourselves and where we're being led. The question is, what are we to do about these issues? Are they real or are they overblown, and what realistic action can we take to avert them?

In 1660, Blaise Pascal, the mathematician who invented probability theory, posed himself the question: 'God is or God is not – which way should we incline?' His answer was that we shouldn't look at the first part of the question – whether or not God exists – for that we couldn't know.

Instead, we must look at the second part of the question – how should we behave? Pascal compared the consequences of behaving as if God didn't exist with behaving as if God exists. If you think God doesn't exist, then your natural inclination might be to go out, drink and carouse, and generally behave badly. If you think God does exist, you will be more inclined to stay home quietly and pray. In either case, there's a chance you might be wrong, but the consequences of behaving as if God does exist (being a bit dull and bored throughout your worldly life) are somewhat more bearable than the consequences of behaving as he doesn't and being wrong (roasting over the fires of hell for eternity). Best really to behave as if he does exist and take your chances.

In a way, the same is true of Crunch Time science. Martin Rees' 50:50 assessment of our chances notwithstanding, we can't know if humankind is laying the foundations for its own extinction, but we can behave as if that's the case. Even if we could dismiss the dark visions of doomsters such as Joy, Kurzweil, Moravec, Rees and the rest as science-fiction fantasy, there are still strong reasons to be concerned about the spread of new science and technology. The spectre of terrorists using weapons constructed from biotechnology, information or nuclear science; berserk nanobots wreaking havoc upon the world we live in; uncontrollable human or computer-created virus pandemics – these things are all real Crunch Time possibilities. What's more, each one of them presents such a vast downside risk (up to and including extinction) that even the most cautious observer would have to agree that it would be wiser to, somehow, restrict and control the progress of science than to allow it to proceed unchecked.

The problem, as with all the Crunch Time issues we've

discussed, is one of balance. The key difference between nuclear technology and much of the science being pursued today is that, in isolation, each of the GRIN technologies is being pursued for peaceful applications – to aid in the human struggle against disease and ageing, to expand our ability to build and create, to continue the never-ending push against the bounds of the human condition. But together they pose a real threat. Even though nuclear technology has since been used (albeit cynically and rather inefficiently) for energy creation and for medical research, the threats presented by nuclear weapons easily overshadow their peaceful use.

Some say this isn't true for the GRIN technologies. While governments with big budgets and huge research facilities pursue nuclear technologies, less ambitious science is practised all around the world in different ways and with differing goals. Each individual advance in genetics or robotics might be driven by the scientist's need to know, but together they hold out the promise of truly transforming civilisation. They too pose a threat, but also immense promise for many Crunch Time issues of poverty, hunger and disease.

For the man who started this debate, Bill Joy, the answer is to 'just say no'. Scientists, says Joy, should give up pursuing technologies that could pose a danger to humankind. He reckons governments should ban pursuit of potentially dangerous science, and impose a regime of inspection and verification. As an example of voluntary relinquishment, he cites the United States' unilateral signing of conventions against biological and chemical weapons in 1972 and 1993 respectively. The challenge, says Joy, will be to apply the philosophy of relinquishment to technologies that are more commercial in application than military.

Joy isn't alone in calling for caution across the scientific world. According to Martin Rees: 'The surest safeguard against a new danger would be to deny the world the basic science that underpins it.' He writes, 'should support be withdrawn from a line of "pure" research, even if it is undeniably interesting, if there is reason to expect that the outcome will be misused? I think it should.'

Will relinquishment be possible? You must be kidding. There's unlikely to be any great global agreement on what's to be restricted and what's not – if one country decides to block scientific investigation into a particular area, others are only too happy to leap into the newly vacated space. Take stem cell research, for instance. Stem cells are the building blocks from which we all grow. They develop to become foetuses, multiplying and changing into each of the separate parts that make up a human being. All the genetic information needed to create a whole person is contained within them. By growing stem cells in laboratories, scientists hope to find out all manner of things about human biology, growth and development, genetics, disease and the like. Each step forward in stem cell research holds immense promise for medicine, and so attracts enormous public and corporate support. At the same time, opponents fear the implications of providing the power over life and death to scientists and their employers, and point out genuine legal, religious and policy questions arising from experimentation on 'human' material.

While Britain and Singapore actively promote biotech in all its forms, pouring government cash into research, the Australian and United States governments take a conservative view of the 'rights' of foetuses. Even today, the United States suffers from a biotech brain drain as biotechnologists shift countries looking for more enticing research environments, or looser legal arrangements. Unless global agreement can be won on which technologies are safe to pursue and which aren't, local bans on technology will be ineffective, and global agreement is one of the most elusive of Crunch Time aspirations.

Besides, scientific progress doesn't happen in straight lines. Newton's discovery of gravity didn't happen because of a government-funded programme set up to drop apples on his head at regular intervals. It just happened. Funding plus research does not equal discovery. And the application of scientific advances is unpredictable. Take a recent discovery, buckyballs – short for buckminsterfullerene. These tiny, football-shaped carbon molecules were discovered in 1985 by accident, off the back of other separately funded projects. But their discovery won a couple of chemists the Nobel prize and buckyballs are now being considered in applications as diverse as AIDS cures and rocket fuel.

It's a vain hope to think that we can restrict progress through national regulation. Instead, Bill Joy's idea was to appeal to scientists themselves, saying that they should adopt an ethical code of practice, a Hippocratic oath of sorts, that would preclude them from pursuing potentially dangerous lines of inquiry. Scientists must be taught the importance of restraint and be empowered to blow the whistle on their colleagues when necessary. This would answer the call – 50 years after Hiroshima – by the Nobel laureate Hans Bethe, one of the most senior of the surviving members of the Manhattan Project, that all scientists 'cease and desist from work creating, developing, improving, and manufacturing nuclear weapons and other weapons of potential mass destruction'.

The sentiments are admirable; the problem is that relying on personal judgement just isn't good enough. Many of the highly ethical scientists in New Mexico managed to convince themselves that atomic bombs were a necessary evil. And the development of nuclear weaponry found sufficient new moral imperatives after 1945 to keep scientists' consciences clear enough – and this was in a liberal democracy. The same kinds of delicate moral choices weren't so easy to broach for scientists working in authoritarian states. How easy is it for scientists to convince themselves that a particular technology is being pursued for good reasons, only to have their research subverted or repurposed later by someone else?

Personal morality, professional ethics – these aren't unimportant things. But we know they can't be effective gatekeepers on a huge scientific population. An appeal to the individual scientific conscience may well pick off the odd doubter, but as an effective means of controlling the future path of knowledge, it's a non-starter.

So if national regulation and personal politics aren't enough, what about global institutions? Unlikely. Take for instance the International Atomic Energy Authority (IAEA). It was set up half a century ago by the UN to check the safety of nuclear reactors in member states, and to make sure they didn't use civilian power plants to kick-start nuclear weapons programmes. But if the IAEA was such an effective body, why has the US drawn up plans to bomb Iran's nuclear plants? Unlike the US, the IAEA doesn't have the power to go in and force countries to do what they said they'd do. International treaties are just too difficult to enforce. When the United States signed up for the 1972 biological weapons treaty, the Soviet Union actually increased its research into anthrax and other bioweaponry.

There's no global police force empowered to check compliance with international agreements and penalise violators – and no such body is likely to appear in the immediate future. The closest we have is the United States of America and many would say it is as much a part of the problem as the solution.

The world's only remaining superpower resists attempts by any outside body to regulate its own affairs. Just one scientific example (never mind about global warming or international law), the United States government opposes any strengthening of the international protocols against biological weapons because letting international inspectors into their facilities would compromise not only America's security, but its 'competitive advantage'.

The answer, perhaps, may be to worry less about the technology itself and more about the health of the societies in which they are embedded. For James Hughes, a professor of public policy (and secretary of the Transhumanist Association), the answer lies in the health of our democracies. Only in a culture in which openness and transparency are fundamental principles can we be as sure as we can that dangers are being risk managed. He sees Joy's call for relinquishment as a kind of Luddism – a 'misplaced attack on technologies that should more properly be directed at the unaccountable powers that deploy them'.

Hughes points out that in authoritarian states, people aren't in a position to force regulatory agencies to do their jobs. This is as true for science as for environmental, economic and other fields that pose risks to our future. Citizens of Communist states weren't able to protest about the spread of nuclear weapons, or to shape policy, just as they couldn't pressure the authorities to ensure correct safety standards at Chernobyl.

Hughes says, 'the next necessary steps in preparing for apocalyptic technologies is to ensure that all societies are open, guaranteeing the rights to investigate, organize and pressure for public health and safety, and that citizens are organized to counter corporate and military domination of the national and global state'.

Rather than call for a halt in scientific progress, a call doomed to failure, the solution is to strengthen civil society. Overly optimistic, perhaps, but civil society, after all, is just the sum of you, me and the rest of us, and our willingness to engage with the wider world.

Breakthrough thinking

It was the middle of a heatwave in the UK, and Adrian was sweltering in 35°C heat on the lawn in Kent. Mike, in Bondi, huddled in his shed on a cool winter's night. The telephone, however, was running hot.

'You don't actually believe all this rubbish, do you?' Adrian was saying. 'About robots gone mad and genetically engineered plagues let loose on an unsuspecting population? It's all a bit Hollywood.'

There was a long pause at the end of the line.

'It doesn't really matter if I believe it or not', came the response. 'What matters is that I'm not really in a position to understand whether it's a possibility, or even to know that there's somebody out there whose job it is to know whether it's a possibility or not. It comes down to the fact that I feel helpless against greater forces. I don't feel like I have a say in the future, and I feel like I should.'

People have never had a say in the future, replied Adrian. That's not what life is about. When people felt helpless in years gone by, they prayed: in fact, they still do.

Societies used to bend their efforts to building ever-greater churches, cathedrals and temples to future-proof them against all the nasty things that might happen. But that isn't enough for us anymore, now we want to know that every risk we take is assessed and managed and controlled.

If something goes wrong, we need to blame somebody in particular, some regulatory body for not doing its job, or some politician for not implementing the right policy. Perhaps it was just easier when we could turn around and say we should have sacrificed a few more chickens.

Mike sighed. 'Like you said, the point is that it's not enough for us these days.'

What society has to do

> *The Industrial Revolution and its consequences have been a disaster for the human race.*
>
> Unabomber/Theodore Kaczynski, *The Unabomber Manifesto*

The march of science might seem unstoppable, thought the Unabomber, but a few well-placed munitions might convince scientists to down tools en masse. During a seventeen-year terror campaign sending mail bombs, he killed three people and injured many more.

The Unabomber was hoisted by his own petard. Both the *New York Times* and the *Washington Post* published his 35,000-word manifesto, after he claimed the killing would stop if they did so. His brother recognised the rambling prose style and shopped him to the FBI.

Clearly, it's going to take a bit more than a few parcel bombs to stop the onward march of science.

So, what would be enough? The point of this chapter

isn't to scare people witless about the terrors that technology will inevitably bring, or to get them out onto the streets with placards reading 'Down With Nanotech', or 'Stem Cell Research Kills Babies'. Instead, it's to ensure that we understand the risks and rewards of scientific progress, and to make clear that it's possible to control and manage those risks.

We need communities that are comfortable talking about the real issues – that aren't overwhelmed by special interest groups, or by the overly conservative or the overly libertarian. Our 21st-century societies need to be able to take decisions based on reasonable information and rational principles backed up by dollops of human instinct, with citizens who understand the issues, care about them, and are empowered to participate in the debate and shape its outcome through the important parts of the policy formulation process. It's only if we fail in this that the more gruesome sci-fi spectres may come to pass.

It's unlikely there will ever be a global authority that can protect us from weird science. Once again, Crunch Time has laid the issue at our door. For that reason it's essential that we understand the challenges that scientific progress presents to us, and that we're prepared to act.

The task is big, but far from hopeless. It's not necessary to be an astrophysicist or a biochemist to understand the issues behind scientific research. You don't have to understand science to understand that it's wrong that 90 per cent of the £40 billion spent on the world's health research is devoted to diseases that account for less than a tenth of the global disease burden, problems like male pattern baldness and acne.

Science has given us access to information and tools to make ourselves heard. A surf on the internet to gather information on potential diagnoses precedes a visit to the doctor. Lobbying by ethical groups is organised by email and mobile phone. The fundamentals, however, are energy and engagement.

Einstein may have said it best in 1954.

> A human being is part of a whole called by us the 'Universe', a part limited in time and space. He experiences himself, his thoughts and feelings, as something separated from the rest – a kind of optical delusion of his consciousness. This delusion is a kind of prison for us ... Our task must be to free ourselves from this prison by widening our circles of compassion to embrace all living creatures and the whole of nature in its beauty.

Yes, even Einstein could come over all mushy when the occasion demanded.

More than half a century has passed since Oppenheimer and his team invented the atomic bomb. It hasn't always been comfortable or secure (and we still might be just seven minutes from midnight), but so far we've managed not to blow ourselves entirely to buggery. Eternity is a long time, but we started it back in New Mexico.

Technology has brought us revolutions before. In one earlier technological wave of change, a new world was opened where people could immerse themselves in a form of communication that broke the verbal and physical world of human contact. A world where voices in your head replaced the spoken word and people became able to immerse themselves in a brave new world of ideas, imagination and knowledge. That technology was print, and the information revolution of the 16th century was the book. Literature did change the world. They burned books in the middle of Florence more than half a millennium ago, but books haven't yet brought about the downfall of humankind. Maybe just give them a little more time.

CONCLUSION

Postponing Extinction

> *A man's life, I reflected, is too long a span today for the pace of change. If he lives more than a half century, his familiar world, the world of his youth, fails him like a horse dying under its rider, and he finds himself dealing with a new one which is not really his. A curious contradiction, this: that as medicine prolongs man's span of life, the headlong pace of technological change tends to deprive him, at an earlier age than was ever before the case, of the only world he understands and the only one to which he can be fully oriented.*
>
> George F. Kennan, 'Vignettes from a Tragic Century', 1997

Extinction has a grim feel to it. Biologists tell us it's a fact for 98 per cent of all species. The fossil record confirms the disappearance of all hominin species bar one – us. Although life expectancy has risen, we can nonetheless extrapolate forward with some certainty to a future without us, our children or any of our genetic descendants. The eventual death of our planet, our sun, our galaxy and universe is guaranteed. Still, this doesn't lead back to the simple advice to live every day as if it were your last.

We cast an eye over near history and tell ourselves that we're different. We're amused by the self-delusions and contradictions of previous civilisations. By the Ancient

Athenians, who thought themselves democratic while propping up their city-state with slavery. By the Roman Empire, with its under-floor heating and poisonous lead piping, combining sophistication, violence and decadence. It's sometimes difficult to see that societies, which in retrospect look so precarious and inclined to destruction, seemed robust and strong to those who lived within them. We tell ourselves that trade and technology make us different, but both stretch back into human prehistory.

Our lifestyle currently occupies us so much that we raise it above the value of new generations. That may not result in the end of humanity, but it does challenge what it means to be human. And that challenge won't recede.

In the introduction we quoted Bobby Kennedy. 'Few will have the greatness to bend history, but each of us can work to change a small portion of events, and in the total of all those acts will be written the history of this generation.'

It isn't enough. Our daily lives comprise many units of action, and many transactions, few with an in-built period of reflection. Our collective enterprises cloud responsibility and allow us to organise activities the consequences of which may shame or repel us.

The brakes

Eventually we will run out of the necessary resources to sustain this speed of growth, and the process of change will slow to a halt. The bacterial colony on the Petri dish will run out of nutrients. World population will run out of land surface, perhaps when, in the year 2500, there are 40 trillion people on earth with just a square yard each.

We've come to learn that the world really doesn't work

like this. Our reductionist, atomic, predictable perspective on the world, the rational, scientific view that so dominated the 20th century has itself been eroded. Change is also not predictable and deterministic but, instead, random. The real world doesn't rattle along on the tramlines of classical thinkers like Newton or Marx. We've learned that even the simplest systems in nature tend to behave in a far more complex and unpredictable fashion. They follow a process of development that's chaotic. While the early stages of change may be linear or exponential at later stages, change often occurs in much more dramatic and unpredictable ways.

In a chaotic view of the world, systems become unstable and undergo dramatic change to create new levels of order and complexity. Witness the complex behaviour of water as atmospheric vapour – the clouds above us – or as a liquid in the complexity of rivers, seas and oceans. There's complexity too in the way people gather and disperse in crowds, the way societies change to reflect the values of the individuals within them, the tipping point of fashion, the epidemiology of disease. And this chaotic view of change has major implications for the world in which we live.

Change isn't simple and gradual and linear. Instead it's characterised by disruptions – non-linearities to mathematicians – that lead to complex behaviour. And that complexity frequently produces dramatic rather than gradual change, revolution rather than evolution.

The 'butterfly effect' put in human terms means dramatic change is frequently triggered by a few extraordinary people with extraordinary ideas – be it the non-violent protests of Gandhi's anti-colonialism to Hitler's demonic anti-Semitism. It also means that small gestures on the part of one individual or community can have big consequences.

The most difficult aspect of many of these issues and risks is that their causes are deeply embedded in the way we live. They can't be isolated and extracted from the everyday actions of everyday people. If the simple act of driving my 4x4 to the airport and flying to my best friend's wedding is a direct contributor to rising temperatures across the face of the earth, searching for solutions means little short of a root-and-branch analysis of what it means to be a functioning member of 21st-century society.

Meanwhile, in searching for solutions, even if we knew everything there was to know about one specialisation, that wouldn't be enough. Poverty might be the result of poor economics, but it's also the result of different types of social interactions, cultural issues or even physical processes like transport, pollution, education and health. That's why so many of the sciences are at loggerheads with each other – look at the debates over the environment and economics, free trade and globalisation, security and freedom. Each of these, and of the rest, is the result of serious disagreements between educated people with valid perspectives on a tangled world.

Our brains allow us to seek rewards and avoid costs by making plans. We share these plans through language and example. When we can't get what we want on our own, we get it by exchanging things with other human beings. So we swap rewards like food, security and information. When people can't supply the reward, like curing cancer or restoring the dead, we look for someone to exchange with who can – spirits, gods, shopping.

Back to *Civilization*

So we're entering a new phase of human evolution that's fundamentally different from all that's come before: we're victims of the Chinese curse. In Sid Meier's *Civilization* terms, just as players move from the Iron Age to the Middle Ages by developing construction and cartography, so we are moving from the Post-Modern to the Post-Materialist era by industrialising science, and technologising communications and information.

As players in the great game of *Civilization*, it's the decisions that we make at this juncture, this turn, this roll of the dice that will determine whether we make it through to the next stage of the game; whether we have the food, water, shelter, resources, technology and sense of purpose to see our species through to the next stage of evolution.

As human beings, we have the ability to plan for the future, to imbue actions with meaning and to postpone gratification. At the same time, the human condition drives us to get the most out of our limited lives, to chase happiness in the short term, to do what is best for ourselves here, right now.

For millennia, our ancestors had only basic hopes of those that came after them – shelter, food and a chance to bring forth another generation. We have long outrun those modest evolutionary goals. We've sought to limit our murderous and violent impulses to allow us to live together and prosper. Now we need to stop indulging our appetite for consuming things and develop our appetite for consuming ideas and information.

We're all micro-players in this macro-decision moment. No one individual can do the work for us. And there's no games designer to allow us to reboot and start again. On the other hand, there's certainty in the knowledge that we can't delegate, that every individual has the potential to disrupt the future, and in disrupting it, change it, hopefully for the better.

INDEX

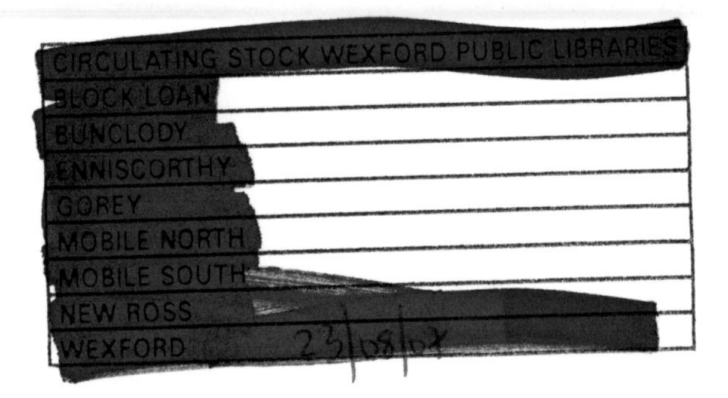